Ralph A. Riccio

THE IRISH ARTILLERY CORPS SINCE 1922

Dedication

To the Gunners, who always serve the guns.

STRATUS

Photo Credits

Philip Bedford	Cavan & Leitrim Railway, Dromod, Co. Leitrim
Broe Collection	Courtesy Janis Pierce Broe and Ray Broe, Dublin
Frank Donaldson	Donaldson Tourism and Heritage Consultancy, Inniscarra, Co. Cork
Bryan Ely	Thurles, Co. Tipperary
Lt Col Edmond Fogarty	Commandant Artillery School, CSC, DFTC, Curragh Camp,Co. Kildare
Jim Horgan	Curator, Collins Barracks Museum, Cork, Co. Cork
Lar Joye	Curator, Art & Industrial Division, National Museum of Ireland, Collins Barracks, Dublin
Roy Kinsella	Cabinteely, Dublin
Brian Lowen	Liverpool, England
MA	Military Archives, Cathal Brugha Barracks, Rathmines, Dublin
Mark McLoughlin	Kildare
David Magee	Fort Dunree Museum, Co. Donegal
Paul Murphy	Clonmel, Co. Tipperary
Sean Murphy	Cork, Co. Cork
National Library of Ireland	National Photographic Archive, Temple Bar, Dublin
Capt Ian Stewart	BK, 2 Bty, 62 Reserve Artillery Regiment, Connolly Barracks, DFTC, The Curragh, Co. Kildare
Colin Stone	Kilkea, Athy
Sean O'Sullivan	Hospital, Co. Limerick

Published in Poland in 2012
by STRATUS s.c.
Po. Box 123,
27-600 Sandomierz 1, Poland
e-mail: office@mmpbooks.biz
for
Mushroom Model Publications,
3 Gloucester Close,
Petersfield,
Hampshire GU32 3AX
e-mail: rogerw@mmpbooks.biz

ISBN
978-83-61421-52-8

Editor in chief
Roger Wallsgrove

Editorial Team
Bartłomiej Belcarz
Robert Pęczkowski
Artur Juszczak

Scale plans
Rodolfo Ciuffoletti

DTP
Bartłomiej Belcarz

Printed by
Drukarnia Diecezjalna,
ul. Żeromskiego 4,
27-600 Sandomierz
www.wds.pl
marketing@wds.pl

PRINTED IN POLAND

Table of contents

Introduction

During the course of research for an earlier book of mine on armoured vehicles in Irish service since 1922, inevitably I ran across references to the use of artillery by the National Army during the Civil War in 1922. Although at the time I paid little heed to the information as I was focused entirely on the use of armour during the war, once research on the armour project had been largely completed, the thought occurred that perhaps a similar book on artillery in Irish service might be a worthwhile endeavour. As I began delving into the subject, it quickly became apparent that there was an overall paucity of information devoted to Irish artillery. Whereas with respect to armour in Ireland there was at least one book that covered the subject to a respectable degree (*Irish Army Vehicles: Transport & Armour Since 1922* by Karl Martin), with the possible exception of the very informative Artillery Corps 75th anniversary commemorative booklet of November 1998, and the earlier November 1973 edition of *An Cosantoir* commemorating the 50th anniversary of the Artillery Corps, nothing of a similar nature existed for the artillery. Because of the important role played by the few artillery pieces in the National Army inventory during the Civil War, a fairly good picture of the use of artillery during that period can be pieced together from a wide variety of sources that deal with the war in a broader context, but from 1923 onwards, except for a few regimental histories varying in detail found during the course of a research visit to the Military Archives in Dublin, precious little exists in any comprehensive or organised fashion concerning the Irish Artillery Corps.

This book is an attempt to present the story of the Irish Artillery Corps since 1922, addressing its history, equipment, and organisation based on all of the information the author has been able to accumulate or access. Despite research at the Military Archives, notable gaps remain, especially with respect to organisation of the Corps over the years and unit equipment issue during The Emergency. One would think that basic details such as date of establishment or disestablishment of at least regimental and perhaps battery size units would be well documented somewhere, but such documentation has proven to be elusive, if not in fact nonexistent. Determining the dates of service and numbers acquired of some of the guns proved to be especially vexing. Records at the British National Archives in Kew likely contain precise information on the numbers and types of guns supplied by the British, the dates they were delivered, and the cost; but efforts to locate them via online searches, but without actually visiting the Archives, were inconclusive.

Perhaps because following the end of the Civil War in 1923 the instances of operational use of artillery by Irish forces has been limited to less than a handful of occasions, the most notable being the reaction by anti-aircraft batteries around Dublin during the 31 May 1941 bombing of that city by the *Luftwaffe*, and the limited use of mortars by the Irish UN contingents in the Congo (1962-1963) Cyprus (1964-1972; deployed but not employed operationally) and Lebanon (1978-2001), the story of the Artillery Corps has not received the type of attention that it merits. Some readers might notice that there is heavier coverage, especially photographically, of 2nd FAR, 6th FAR, 7th FAR, 62 RAR, and 14th, 19th, and 42nd Batteries. This is due to the fact that a few of the individuals mentioned in the acknowledgements below are members of, or had past association with, those units and had extensive photo collections, or had access to images relating to those units.

As this book is being prepared to go to press, it appears that the Irish Government has made a decision to decrease the number of brigades in the Irish Army from three to two; there likely will be a radical reduction and restructuring of the Reserve forces as well. What effect this will have on the structure of the Artillery Corps, or which units will be consolidated or disbanded, has not been determined as of this writing.

It is my hope that, as incomplete as this work may be, it at least begins to tell the story of the artillery in Ireland since 1922 in a somewhat coherent and comprehensive fashion. Although I have tried to verify the accuracy of all of the information presented in this work, in all too many cases I had no choice but to make judgment calls based on conflicting information at hand. In those cases, I have either footnoted or explained in the text itself that there are some doubts as to the total veracity of the information. However, despite my best efforts to ensure accuracy, it would not be surprising that there still may be errors of fact. I would be grateful if any readers who might take issue with anything contained in this book would contact me through the publisher.

Acknowledgements

In some respects I feel that rather than writing this book myself, I have only orchestrated the efforts of a great number of people in several countries to come up with a somewhat coherent history of the Irish Artillery Corps. Among those who deserve special recognition are:

Lt Col Edmond Fogarty, School Commandant Artillery School, CSC DFTC, who took an active interest in this project early on and was not only most generous in providing photographic documentation, but was extremely patient in answering the multitude of sometimes arcane questions that I asked. It was truly comforting to have someone with his experience and depth of knowledge to comment on my work

Denis McCarthy, assuredly the most knowledgeable individual concerning the history of armoured vehicles and artillery in Irish service, who likely has forgotten more about the subject of artillery in Ireland than I have managed to set forth in this volume. He is passionately dedicated to continuing research on Irish military history, and has written countless monographs on equipment used by the Irish Army. Denis provided otherwise unobtainable photographs of various artillery pieces in Irish service, and details of their employment.

Colin Stone and his uncle, Shay Stone, who between them have over sixty years of service with the Artillery Corps, and who were unbelievably generous in sharing their knowledge of all things related to the Corps. Colin initially volunteered to provide some excellent and unusual images from his vast collection of artillery related photos, accompanied by extensive notes and explanations. This was soon followed by an incredibly intense and prolonged search on his part for additional photographs from museums and artillery barracks, resulting in what seemed to be a never ending deluge of photographs. His astounding energy level and ability to find even more photographs seemed to escalate as the book progressed. For his part, Shay was instrumental in providing artillery unit historical data and vintage photographs that I had been unable to obtain elsewhere. Both Colin and Shay were more patient than could have been reasonably expected in response to my requests for specific information on images they provided; their inputs added significantly to the quality and completeness of this book. Colin also managed to drag RSM Mick Dillon into the equation, who provided access to many photographs relating to 2nd FAR, as well as providing some very interesting photographs of equipment in Chad. It would be more than fair to say that without Colin's contribution, especially with respect to the number and quality of photographs he provided, this book would tell a much less complete story of the Irish Artillery Corps.

Comdt Sean Murphy, who at the very outset of this project provided encouragement, as well as furnishing the first photographs and reference material relating to the project, and Comdt Paschal Cullen (Retired), who in turn provided additional photographs and priceless insights to Sean to be passed on to me.

Philip Baldock, the Manager of Newhaven Fort, Sussex, England, for his enthusiastic assistance and great detective work with respect to ex-Ireland coast artillery guns that sound their way back to England and are now at Newhaven Fort.

Capt Ian Stewart, Battery Captain, 2 Bty 62 Reserve Artillery Regiment, Defence Forces Training Centre, The Curragh, Co. Kildare, who quickly integrated himself into the research effort with respect to regimental and battery histories, and who provided a goldmine of photographs, especially of the 6th Artillery Regiment (FCA). His patience in explaining details relative to the photographs is worthy of special thanks.

David Magee, Curator of the Fort Dunree Museum, who schooled me on various aspects of Irish coast defence artillery and who provided photographs of the guns at Dunree.

Roy Kinsella, who spent untold hours tracking down locations associated with gun and anti-aircraft positions around Dublin during The Emergency and photographing sites as they appear today, as well as for photographs of current equipment.

Paul Murphy, for his constant encouragement for yet another book on the Irish military, as well as for his insights and tidbits of background information.

Brian Lowen for some superb vintage artillery photographs from his family collection, and for his useful background information on the 2nd Artillery Regiment.

Bryan Ely, who provided historical reference material, photographs, and personal insights on Reserve Artillery.

Capt Padraic Kennedy, and his successor, Capt Rory Esler, Public Relations Section, Defence Forces Headquarters, for their prompt attention to my many requests for permission to use Defence Forces photographs, as well as to Sgt David Nagle, Editor, *An Cosantoir*, for his help with photographs of current equipment.

Sean O'Sullivan, who to my amazement always seemed to be able to reach into his files to produce photographs of pieces of equipment that I otherwise lacked.

Flt Sgt James Perkins, Irish Air Corps, the justly acknowledged expert on Irish military insignia, who has most generously provided the insignia relative to Irish artillery units that form an important appendix to this book.

Rodolfo Ciuffoletti, who agreed to accept the challenge of creating groundbreaking line drawings for most of the guns covered in this book, and who succeeded admirably, sometimes working with just the sketchiest of documentation.

Roger Wallsgrove, my editor, who again has encouraged this ongoing fascination of mine with Irish military history and has seen fit to publish this companion volume to my earlier volume on AFVs in Irish service, and to Robert Pęczkowski and the staff at Stratus in Poland who have made this book a physical reality.

My wife Charlene who with infinite patience and an unerring eye plodded through the tedious task of proofreading the text.

Thanks and gratitude are due also to the following individuals and organisations, without whose assistance this book would be much less interesting and accurate: Rita Bartholomew; Derek Barton; Brian Baxter (REME); Philip Bedford (Cavan & Leitrim Railway, Dromod, County Leitrim); Andrew Berg; Frank Broaders (Duncannon Fort Vistor Centre, Duncannon, New Ross, Co. Wexford); Janis Pierce Broe and Ray Broe; Glen R. Broman; Rodolfo Ciuffoletti; Deirdre Condon; Sgt Eric Craig (4 FAR, Curator, Mullingar Barracks Museum); Dr. Boyd Dastrup (Command Historian, U.S. Army Artillery School, Fort Sill, Oklahoma); Com Deasy and Robert Montgomery (Mitsubishi Motors Ireland, Dublin); Donard de Cogan; Thomas Diffley, (Westward Scania, Strokestown, County Rosscommon); Frank Donaldson (Donaldson Tourism and Heritage Consultancy, Inniscarra, Co. Cork); Bryan Ely; David Fletcher (Tank Museum Bovington); Martin Forde; Liz Gillis (Kilmainham Archive); Jim Horgan (Curator, Military Museum Collins Barracks, Cork); Lar Joye (Curator, Art & Industrial Division, National Museum of Ireland, Collins Barracks, Dublin); Anne Kearney (Irish Examiner); Comdt Victor Laing, Hugh Beckett (Military Archives); Patrick Lynch; Donal MacCarron; Paddy McGannon; Mrs. Pamela Manahan, Sandycove, Co. Dublin; Sandra McDermott, Anne O'Donoghue (National Library of Ireland); Paul McDonald; Mark McLoughlin; Paul McMenamin; Karen Millward; BS Frank O'Connor (2 FAR); Gerry O'Connor; Sharon O'Donovan (Mercier Press); Lt Col Ciaran O'Halloran (Retired); Cormac O'Keeffe; BQM Gerry O'Reilly; Sgt Damien Quinn (HQ 4th Field Artillery Regiment); Liam Reilly (Millmount Museum, Drogheda); Comdt Sean Reilly (Bty Cdr, 3rd Bty, 62 RAR); Sgt Mick Royanne (1 FAR), Chris Sanders, Early Ford V8 Club, UK; Dawn Steele (Lancashire DAF, Bamber Bridge, Preston, Lancashire); Ed Tyhurst (Newhaven Fort, Sussex, England); James K. Wagner, Early Ford V8 Club of America; John Walsh (Bere Island Projects Group); Tony Wade, (Battery Quartermaster, 54 Reserve Artillery Regiment); Darren Waters; Terry Ward; Gabriel Whitson, Gabriele Zenoni.

Artillery During the Civil War (1922)

The Irish Artillery Corps emblem, with the inscription in Gaelic.

The birth of what is now the Irish Artillery Corps can be traced to Tuesday evening, 27 June 1922, when a group of four National Army officers from Beggar's Bush Barracks in Dublin arrived at the British-held Marlborough Barracks (now McKee Barracks) in Phoenix Park to collect two 18-pdr field guns and ammunition. The transfer had been arranged by General Eoin O'Duffy, Assistant Chief of Staff of the National Army, in communication with General Sir Nevil Macready, General Officer Commanding, British forces in Ireland. The four National Army officers sent to collect the guns were General Emmet Dalton, Colonel Tony Lawlor, Captain Peadar McMahon, and Captain John C.F. Doyle. The two guns, having been relinquished by the British, were brought by motor lorries to Winetavern Street on the south side of the River Liffey, and St. Mary's Abbey on the north side of the river.

There was hardly a pause between the receipt of the guns and their baptism of fire in Irish service – indeed, it was only a matter of hours. The first use of artillery by Irish forces can be dated fairly precisely, almost to the minute. Accounts by various eyewitnesses state that the first shot was fired by Irishman against Irishman sometime between 4:07 and 4:29 a.m. on the morning of Wednesday, 28 June 1922, when National Army gunners fired against the Four Courts building complex occupied by Republican forces. This twenty-two minute discrepancy may be attributed to watches or clocks set at different times, or to some inaccuracy in observation or recollection, given the excitement of the moment. What is much more certain is that the first round was fired by a Mk II 18-pdr, serial number 10756, positioned at the corner of Winetavern Street and Merchant's Quay, on the south side of the River Liffey across from the Four Courts on Inns Quay. The gun crew consisted of Colonel A.T. "Tony" Lawlor in charge of the gun, as well as Captain Johnny Doyle, a Corporal McLauglin, and two privates.[1] It is ironic that this inauguration of Irish artillery should take place in Dublin and be directed by Irishmen against their fellow Irishmen. The other gun,

1. *The gun reportedly fired a total of 375 shells at the Four Courts complex between 28 and 30 June, when the Four Courts complex suffered catastrophic damage from an explosion inside the buildings themselves.*

One of the early 18-pdrs provided to the National Army by the British in June 1922. The name on the upper shield, "Hammond Lane No 4" may be a reference to the scrap merchant who has long been and is still engaged in disposing of the Irish Army's scrap. It is being towed by a heavy lorry, which lends some credence to the supposition that it may have been one of the two original guns supplied on 27 June 1922. (NLI INDH 242, Courtesy of the National Library of Ireland)

under Peadar McMahon's command, was retained in reserve at St. Mary's Abbey, on the north side of the river and northeast of the Four Courts complex.

The National Army troops attacking the Four Courts were under the overall command of General Tom Ennis, but General Emmet Dalton was in charge of the two 18-pdr field guns that had been handed over by the British. The Republicans claimed that the gun crews were in fact British personnel, or that at least the officers were Royal Artillery officers. Neither of these claims was true, although some sources suggest that some of the gunners had served in the British Army. Michael Collins himself is said to have urgently sought gunners with World War I experience while the shelling was taking place. Given the somewhat erratic performance of the guns, it does seem improbable that there were any truly qualified artillery personnel on the scene at the time. In fact, at one point Dalton took personal charge of the guns himself, working with Colonel Tony Lawlor and his gunners even though his knowledge of artillery seems to have been limited to sighting along the bore and then firing.

In an attempt to set the record straight with respect to the use of artillery by the National Army during the Dublin fighting, the 3 July 1922, issue of *The Irish Times* carried the following official bulletin issued from General Headquarters at 5:30 on the previous evening:

In view of the fact that fictitious stories are being circulated, it is necessary to state that the artillery used against the Four Courts was under the command of Major-General Dalton, who, it will be remembered, entered Mountjoy Prison in the attempt to rescue Major-General Sean McKeon last year. Major General Dalton had four 18-pounder guns under his control, and these guns were worked in two sections of two guns each. Commandant General Lawlor, with two teams of gunners from General McKeon's division[2] *manned the section which operated on the south side of the river. The other section was divided, and one gun was under the charge of Commandant P. McMahon, whose team was composed of gunners from the Midland Division, and also of gunners from the Curragh Camp. One gun was kept in reserve by Major General Dalton. It is desired to contradict the*

2. *The General McKeon referenced in the report is General Sean MacEion, his division being the Midland Division.*

allegations being made from certain quarters with respect to the use of artillery, and the public will clearly understand that the artillery was controlled, directed, and manned by officers and men of the regular I.R.A. The field pieces were used principally in order to effect breaches in the walls through which our troops could storm the building. No incendiary shells were used, and artillery was not the cause of the conflagration, which was really caused by the land mine exploded by Irregulars of the Four Courts garrison.

The purpose of the bulletin would appear to have been to discredit reports by the Republicans that any of the guns were manned by British personnel, and in a bit of gratuitous irony, referred to the artillery as having been "...controlled., directed, and manned by officers and men of the regular I.R.A.", challenging the claim made by the anti-Treaty forces that they represented the true I.R.A. The details provided in the bulletin also manage somewhat circuitously to shed some light as to the number of guns involved in the Dublin fighting. Although it is certain that initially only two guns were available to Dalton (one each under Lawlor and McMahon), some accounts refer to a total of four guns being used in the fighting. The excellent first-hand account by Niall Harrington states unequivocally that only two guns were available initially, and that the one at St. Mary's Abbey was not brought into action. He further states that the changing of position of the one gun from Winetavern Street to Bridge Street accounts for the belief that two guns shelled the Four Courts from south of the Liffey.[3] An iconic photograph supposedly taken on the morning of 30 June, however, clearly shows two 18-pdrs and two Lancia amoured personnel carriers in position at the corner of Winetavern Street and Merchant's Quay/Wood Quay. The explanation of the apparent contradiction lies simply in the fact that between 28 June and 2 July the British turned over two additional 18-pdrs, bringing the total to four guns. What remains elusive is the number of guns actually committed to action during that five day period: scrutiny of photographs and

3. *Harrington, page 67.*

A Mk I 18-pdr being manhandled to a new position during the fighting in Dublin. Although normally towed by a Lancia troop carrier, the guns could be moved relatively easily by a group of gunners for short distances over the paved or cobbled streets. (Kilmainham Goal Archive)

An 18-pdr in action in Dublin. The direction in which the gun is oriented suggests that it was positioned at the corner of Bridge Street and Merchant's Quay, firing in a northeasterly direction. A Lancia personnel carrier parked to the right of the gun provides some protection to the crew. A number of rounds laid upon wood batting are in the middle of the cobbled street. (NLI HOG 185, Courtesy of the National Library of Ireland)

accounts of the fighting confirms that two guns were in action together at Winetavern Street, probably on 30 June, this being the section under Lawlor's command as described in the 2 July bulletin, but it has proven impossible to determine whether all four guns were committed to action, or if at least one was kept in reserve.

Photographs taken during the fighting in Dublin almost invariably show the presence of Lancia armoured personnel carriers alongside the 18-pdrs. The Lancias were used to shield the gunners from sniper fire and often were used to tow the guns as well. In fact, throughout the Civil War the 18-pdrs were almost exclusively towed by mechanical transport (either by the Lancia armoured carriers or by heavy lorries), or were carried portee-style on flatbed trucks. Based on all of the

One of the two 18-pdrs on the south side of the Liffey, firing at the Four Courts complex from Bridge Street. Damage to the west wing of the complex, directly across from the gun, is evident, suggesting that at least some high explosive shells were available. The wooden crates in the foreground appear to be filled with cobbles from the street, possibly having been torn up to enable the gun spade to be properly anchored. The side of a Lancia armoured carrier can be seen to the right of the gun. (Courtesy of Mercier Archives, Mercier Press Ltd., Cork)

reference material consulted, instances of relying upon draught animals to tow the guns during the Civil War were rare indeed, and were limited to occasions when mechanical transport broke down or was unable to negotiate particularly difficult terrain.

Although the British had provided the guns to the National Army, they do not seem to have provided any training along with the guns, except probably for some extremely rudimentary and perfunctory instructions on how to aim, load, and fire the pieces. Indeed, the party of officers under Dalton that went to pick up the guns at Phoenix Park on the evening of 27 June would have had little time for more than the briefest familiarization with the weapons before hauling them into position to be put into action. The newly minted gunners who manned the guns, seemingly lacking any experience or training on how to sight the guns, turned to the rather simple expedient of sighting along the top of the barrel, or possibly the slightly more reliable method of bore sighting (i.e., with the breech of the gun open, looking down the empty barrel and elevating and traversing until the intended target came in view). Although lacking in any sophistication, these methods proved effective enough, as the target, namely the Four Courts building complex immediately across the River Liffey was close, immense, and immobile. The close range was such that factors normally affecting artillery fire, such as ambient temperature, humidity, and wind speed, would have only minimal, if any, effect on the trajectory of the rounds. There would seem to have been literally no possibility of rounds falling short or over the target, although due to the inexperience of the gunners and the fact that they were being subjected to hostile small-arms and machinegun fire, despite the odds both of these improbabilities did, in fact, occur. One excited or distracted gunner did manage to fire low and blow off a chunk of the river wall on his own side of the river at the Bridge Street position.[4] There is likewise an account of one of the gunners, one Ignatius O'Neill from County Clare, elevating the barrel of an 18-pdr as high as he could in an attempt to hit a sniper in the dome of the Four Courts[5]; reportedly, the shells went through the dome and landed on the grounds of the Royal Hospital some two kilometers away in Kilmainham.

An 18-pdr field gun in action at the intersection of Henry Street and Sackville Street (now O'Connell Street), probably on 4 July. When this photograph was taken, there were only four 18-pdrs in the National Army inventory. Behind the gun is a Lancia armoured troop carrier; the Fianna Fail symbol is partially visible above the gun shield. Armoured Lancias frequently towed the 18-pdrs in Dublin, and then were parked in positions that provided the gunners some protection against small arms fire. (Broe Collection)

The availability of ammunition, or the lack of it, was the cause of some concern; only 20 rounds per gun had been supplied with the first two guns. In order to husband the small stock of ammunition that had been supplied by the British, initially the gun firing from Winetavern Street did not maintain a sustained fire, but rather fired at about five minute intervals. At this early point in the fighting, the National Army forces also held out the hope that the Republicans in the Four Courts would decide that they were surrounded and outgunned and would give up the fight. That, of course, did not prove to be the case. With ammunition stocks precariously low, the Provisional Government sought another 300 rounds from the British, but Sir Nevil Macready agreed only to provide another fifty rounds of shrapnel (allegedly all that was available to him at the time).[6] In

4. *It has also been suggested that the short round was, in fact, a defective round that did not clear the wall's parapet.*

5. *Calton Younger, Ireland's Civil War, page 326. Although Younger's account, as written, would seem to indicate that O'Neill's gun was one of Lawlor's guns on the south side of the Liffey, other information points to the gun being on the north side of the river. In fact, Kilmainham is itself on the south side of the river, so for a shell to land on the grounds of the Royal Hospital, it would have to be fired from the northern bank, traveling south, or more precisely, southwest.*

6. *In view of the fact that the one 18-pdr initially brought into action at Winetavern Street reportedly fired a total of 375 shells during the attack, it is clear that ultimately the British provided far more ammunition than Macready claimed to have available.*

fact, the shortage of ammunition was compounded by the fact that the shells that already had been supplied were shrapnel shells designed to cut and clear wire obstacles, rather than high explosive shells (although Macready maintained that he had provided Dalton with "many" HE shells). As a consequence, the artillery was largely ineffective, as although the shrapnel rounds made much noise when they hit the solid walls of the Four Courts, they did relatively little damage. Urgent demands by the Provisional Government for more ammunition ultimately led Macready to send a destroyer to Carrickfergus Arsenal (in what is now Northern Ireland) to obtain an adequate stock of ammunition to supply the guns that had been provided to the National Army. The turnover of additional 18-pdrs also made it incumbent upon the British to supply ammunition on a scale commensurate with the increased number of guns in the inventory.

Because of sniper fire from the Four Courts (which the aforementioned O'Neill tried to neutralize with the 18-pdr), sometime on 28 June Lawlor moved his gun from Winetavern Street west to Bridge Street, across from the western face of the Four Courts. The artillery continued to fire throughout Wednesday and into Thursday, with the shells finally cracking a wall in the western wing of the complex. On Thursday night, 29 June, Lawlor was ordered by Dalton to move his gun back to Winetavern Street. In the meantime, the British had supplied two more 18-pdrs, one of which joined the first gun on Winetavern Street. Younger refers to guns deployed to Green Street under Peadar McMahon,[7] north of the Four Courts, which would have allowed the National Army forces to bring fire to bear on the Four Courts from both the south and the north of the complex. Although there is virtually no mention in any publications that any of the guns north of the river were brought into action against the Four Courts, it seems highly likely, if indeed not certain, that at least one of the guns under Peadar McMahon was brought into action. According to Lt Col Edmond Fogarty, Commandant of the Artillery School in Curragh Camp, Artillery Corps folklore holds that the gun northeast of the Four Courts missed its target on a number of occasions and that its projectiles landed in the Royal Hospital, Kilmainham (now the Irish Museum of Modern Art), which had become the headquarters of the withdrawing British forces, and that, understandably, the British GOC was "not amused."[8] To add to the embarrassment, reportedly some of the shells landed just as General Emmet Dalton was meeting with General Sir Nevil Macready on the Royal Hospital steps. This gun undoubtedly is the gun referenced earlier by Younger, which Gunner O'Neill was using as a sniper's weapon. In order for shells to have landed on the Royal Hospital grounds, they would have had to have been fired along an axis that ran from northeast to southwest, with point of origin northeast of the Four Courts; plotting such an axis on a map suggests that the gun could have been positioned at the intersection of Mary and Capel Streets or a nearby intersection. At about this time the British had also offered the use of 60-pdr howitzers to compensate for the ineffectiveness of the 18-pdrs against the stout walls of the Four Courts, but the Provisional Government refused the offer. The following day, Friday, 30 June, a series of explosions and fires in the Four Courts, caused by the ignition of munitions stored in the buildings, finally caused the Republican defenders to surrender at 3:30 in the afternoon. That, however, did not mark the end of the fighting in Dublin, as the Republicans sought to make a determined stand on the eastern side of Sackville Street (now O'Connell Street),

The 18-pdr field gun in position on Henry Street during a lull in the fighting, The gun crew have quit their gun for the moment, and some civilians seem to be engaging them in conversation. (Broe Collection)

7. *Calton Younger, page 328. The gun or guns apparently were redeployed from St. Mary's Abbey slightly north and west to Green Street.*
8. *E-mail dated 17 August, 2009, from Lt Col Edmond Fogarty to the author.*

barricading themselves in a number of hotels (among them the Crown, Granville, Gresham, and Hammam) and adjoining buildings that became known as "The Block."

The 18-pdrs were soon in action against stubborn targets on The Block, as well as other Republican strongholds in the area. On Sunday, 2 July, an 18-pdr, reportedly mounted on a flat-bed truck,[9] took a position at Beresford Place and Amiens Street (site of the current Busarus bus terminal) and began to pump shells into Moran's Hotel, at the juncture of Gardiner and Talbot Streets. Moran's had held out against attacks by an armoured car earlier in the day, but after the 18-pdr fired three shells into the building, the defenders quit the site and took shelter in Hughes' Hotel, also in Gardiner Street. On Tuesday, 4 July, an 18-pdr was positioned on Henry Street where it intersected with O'Connell Street and brought its fire to bear on the Granville Hotel, which was a Republican strongpoint. The following day, although the artillery fired only twenty-five rounds during the course of the day, they had a devastating effect. The Hammam Hotel collapsed in flames.

In summary, with respect to the fighting in Dublin at the outset of the Civil War, the four 18-pdr field guns available to the Provisional Government forces conferred a notable edge against the Republicans who had chosen some substantial defensive positions. The Republican forces, however, had no effective means to counter the National Army's artillery. The 18-pdrs could be moved about with relative ease and impunity, and to all intents and purposes were able to engage in near point-blank fire against Republican strongholds. Artillery clearly showed itself to be the "Queen of Battle" during the fighting in Dublin.

The fighting in Dublin had not yet come to an end before the National Army began to conduct operations against Republican positions outside of Dublin. The dust had barely settled from the 30 June destruction of the Four Courts when a decision was made, presumably at GHQ level (likely by Collins or Mulcahy) to release one of the 18-pdrs (in fact, it was the gun that fired the first round of the war from Winetavern Street on 27 June) to deal with some anti-Treaty forces in Drogheda, some 25 miles north of Dublin. The decision could have been made no later than 2 July, or the morning of 3 July, as on 3 July, National Army forces had advanced towards Drogheda, accompanied by the 18-pdr. The Martello tower at Millmount, Drogheda had been occupied by some 35 Irregular troops who, despite an appeal from the parish priest of St. Mary's Church, which was in close proximity to Millmount, refused to leave, no doubt causing some anxiety on the part of the priest who feared damage to his church. The gun remained overnight in a barrack yard in Slane, about seven or eight miles west of Drogheda, but on the morning of 4 July was put in position close to Cahill's bookbinding works on Dominic Street (now Murdock's car park). Although the gun position was on the north side of the River Boyne, in County Louth, and the Martello tower was to the south of the river in County Meath, the distance was only about 1200 meters and there was a clear line of fire. Once begun, the shelling continued for six hours; according to the gun's record sheet, 40 rounds were fired. The tower was seriously damaged, but there were no deaths or injuries, as the anti-Treaty troops had fled, despite their refusal to do so the day before. It seems that the gunners, under Captain Johnny Doyle, still had not learned the finer points of aiming the guns, as the firing was described as quite erratic. The first round, fired as a ranging round, landed some 2000 meters southeast of the target, in a field at Beamore in County Meath. The following rounds closed on the target, causing extensive damage both to the Martello tower as well as to other buildings in the vicinity that happened to be struck by the less than totally accurate fire. The entries for 4 and 5 July in the Police Book of West Gate Barracks, sum up the fighting very succinctly:

4 July: Big gun arrives in Drogheda. Heavy fire from same directed at Millmount Barracks (sic!) Regular forces under Captain Stapleton and Commandant Mooney take Millmount Barracks without any difficulty.

5 July: Big gun leaves town. Position in town quite normal.

---

9. *The author believes that this reference to an 18-pdr on a flatbed truck is evidence of at least a third gun in action during the fighting. Photographs show that the two guns under Lawlor's command are in the company of Lancia personnel carriers, which probably were used to tow the guns as well as to protect the crews from small arms fire. Although speculative, it would seem that mounting a gun on a flatbed would not be necessary if there were already a prime mover dedicated to the gun. This gun may have been one of the guns under Peadar McMahon's command.*

At the same time, National Army forces had begun to advance towards Blessington, south of Dublin, but halted in order to await the arrival of artillery to support the advancing columns. At that date (4 July), the total artillery inventory stood at four guns, but National Army commanders clearly understood the importance that artillery had played in the fighting that still had not ended in Dublin, and the huge advantage it provided against the opposing Republican forces who, entirely bereft of artillery themselves, had no effective defense against the guns. This was demonstrated during the National Army advance on Boyle in County Rosscommon (some ninety miles northwest of Dublin) the following day, 5 July, when an 18-pdr under Sean MacEion arrived on the scene, causing the Republicans to abandon the town and rally at Ballaghaderreen and Charlestown. Precisely which of the four guns was assigned to MacEion is impossible to determine, but it is noteworthy that General Headquarters had no reluctance to assign scarce artillery assets to support operations being undertaken quite a distance from Dublin.

The transfer of additional 18-pdrs by the British continued at a fairly brisk pace, as the following actions that took place between 6 July and 8 August seem to have involved at least six and possibly seven different 18-pdrs. By the end of 1922 a total of nine guns had been handed over by the British; most of the guns were parceled out to individual commands, with a few probably being retained in Dublin to be distributed on an as-needed basis.

On 6 July, a Provisional Government expeditionary force consisting of some 250 officers and men supported by an 18-pdr field piece, two armoured cars, and two Lancia armoured troop carriers set out for Wexford. This is probably the first manifestation of the National Army's consistent use of what today would be referred to as combined arms teams of infantry, armour and artillery; it was a pattern that was repeated in virtually every significant National Army operation of the war. The Republicans, lacking artillery and to a lesser extent the armoured cars available to the government forces, by circumstances as well as by choice, were forced to conduct an almost exclusively defensive strategy. On 7 July, as the advancing column reached Blessington with its artillery, the Republican forces simply withdrew rather than face the prospect of an artillery bombardment. The threat posed by even a single gun caused a similar reaction on several other occasions. By 8 July, the National Army column consisting of 22 trucks, the armoured cars, and the artillery piece was approaching North Wexford through Wicklow.

Meanwhile, far to the north in County Sligo, on 13 July National Army forces under MacEion, continuing their advance north from Boyle in County Rosscommon, were attacking Republican forces in Collooney, Co. Sligo. Anticipating a stout resistance, he brought up his artillery, consisting of the single 18-pdr field piece that had proven so intimidating little more than a week earlier at Boyle. The gun was positioned in a sawmill, and it soon attracted a burst of machinegun fire from the tower of a Protestant church. One of the bullets creased the scalp of one of the artillery crew, a Sergeant Cassidy, who after recovering from his shock became extremely irate at what had happened. He took cover behind the gun's shield and upon determining the location of the machinegun, trained the gun on the church tower. The first round was on target, going through a belfry window and exploding, ringing the bells as well as blowing the top of the tower into the air. Amazingly, the part of the tower that had been lifted into the air fell back more or less to its original position, and miraculously, none of the Republican machinegun crew who had experienced being lifted somewhat closer to heaven by the blast, were wounded.[10] An infantry attack dislodged the defenders.

Less than a week later, on 19 July, an 18-pdr, positioned on Mount Misery, across the River Suir from Waterford itself, was committed to action during the battle for Waterford. Colonel Patrick Paul, a Watefrord native who had served as a gunnery officer in the British Army during World War I, was directly in charge of the attack on his native city. The day was said to be a perfect day for shelling, as visibility was excellent, allowing the fall of the shells to be observed with the naked eye. The gunners used the two towers of Ballybricken Prison as reference points. The gun then began

10. Calton Younger, pages 364-365.

to fire at Republican-held targets, including the Infantry Barracks, Cavalry Barracks, and the jail, all conveniently located in a straight line with respect to the axis of the gun, at a range of 1200 to 2500 yards. Some shrapnel shells were used for registering targets, which, once acquired, were shelled with high explosive shells. Colonel Paul himself directed the fire of the gun, but in a somewhat embarrassing incident, a shell directed at one of the barracks was a near-miss and hit his own house on Newgate Street while his mother was in it.[11] Fortunately, his mother, who was cooking in the kitchen at the time, escaped unscathed. The following day, after seizing a number of buildings, National Army troops were stalled by effective fire from the Post Office. The field gun was brought to the scene and lobbed a few shells at the Post Office, putting an end to trouble from that location. The amount of shelling, and the change in location of the gun, led the Republicans to claim that a full four-gun battery had been brought into action against them, but this simply was not true. One source states that his enquiries lead him to believe that two artillery pieces of different calibre were in action,[12] but this belief seems to be based on the fact that two different types of shell were used in the attack (HE and shrapnel), and ignores the fact that Paul's 18-pdr fired both types of shell during the action. This was not the only questionable claim made by the Republicans with respect to the artillery at Waterford. By 20 July, the Cavalry and Infantry Barracks were being devoured by flames, which the Republicans claimed had been caused by incendiary shells. Colonel Paul in turn stated that no incendiary shells were used, and that only shrapnel and HE shells were used. Although it is possible that hot splinters from HE shells could have been the cause of some fires, given the Republican practice of setting installations on fire prior to abandoning them, it is more likely that most of the fires were deliberately set by the departing defenders.

By 21 July, the Republican defensive lines had shrunk, with Ballybricken Prison serving as the focal point of the defence. Under the direction of Paul's superior, General John T. Prout, the 18-pdr was emplaced on the Ferrybank railway bridge to bring fire to bear on the prison. However, at every attempt by the gunners to dig in the trail of the gun, an intense fire issued from the Republican positions, killing the soldiers attempting to get the gun into battery. At this point a sergeant named Howlett sprang up with a Lewis gun, covering the digging party with short bursts from the Lewis gun; he himself soon was riddled with fire from the jail, but his action had enabled the trail to be properly set, and soon thereafter the gun began to spew its shells, breaching the jail walls.[13] The obstinate defenders, faced with the choice of a continued artillery bombardment or surrender, chose the latter, thus ending the battle for Waterford. Although one source reports the number of rounds fired at Waterford as thirty-six,[14] it is not clear if this refers to the total fired by Paul's gun, or if it refers only to the rounds fired at the barracks and jail and does not account for those fired against the Post Office.

While the attack on Waterford was in progress, some eighty miles to the north-west at the other end of the so-called "Republic of Munster", National Army forces were also pushing an offensive against the key city of Limerick. On 17 July Commandant Dinny Galvin arrived by train in Nenagh with an 18-pdr. The following day saw heavy fighting in Limerick by opposing infantry elements. On Wednesday, 19 July (the same day that the 18-pdr in Waterford began its work), Galvin's 18-pdr had been brought down from Nenagh and was set up on Arthur's Quay, ready to fire at the Republican-held Strand Barracks. The defenders were invited to surrender, but not surprisingly declined the invitation, stating that they would hold the position as long as their ammunition held out. Although accounts differ as to the details of the artillery attack, there seems to be agreement that a total of at least 33 shells were fired to breach the walls of Strand Barracks. One account

11. *Eoin Neeson, The Civil War 1922-23, page 155.*

12. *Jack O'Neill, Waterford, A History, page 39.*

13. *Eoin Neeson, page 159, says the sergeant was unnamed and stood in the center of the bridge; Duggan (A History of The Irish Army), page 90, provides the sergeant's name, and states that he fired from a Lancia armoured carrier. Additionally, in an E-mail dated 13 August, 2008, Paul Murphy, originally from Waterford, stated that in 1979 his grandmother related to him that the wounded man was laid out on the kitchen table of her house on Dock Row, Ferrybank, a few hundred yards from where the gun had been dug in.*

14. *Duggan, page 89.*

claims that 33 shells were fired over several hours and breached the front wall of the barracks and that then the gun was moved to fire on the rear wall of the prison[15], while another account states that 19 shells were fired at the front of the building, followed by 14 shells being fired at the back (the total also being 33 shells).[16] One source also mentions the use of mortars in the attack on Strand barracks, one of the very few times that this type of weapon is mentioned in accounts of the fighting.[17] At any rate, the barracks were captured, and the gun was moved to attack the next target, Carlisle Barracks, which likewise soon was in flames. The overworked gunners were once again pressed into service at midnight, firing at the Ordnance Barracks in Musgrave Street. The Republicans quit the barracks, and setting fire to other strongholds as well, withdrew. The battle for Limerick effectively was ended, once again due largely to the intervention of a single field piece against which the Republicans had no viable defense.

Leaving the Republicans no respite, a National Army expeditionary force left Dublin on 22 July aboard the *Arvonia*, bound for a seaborne landing against Westport in County Mayo in the West of Ireland. The force, with a strength of 400 men, included an armoured car and an 18-pdr, as well as 150 bicycles. Added to the innovative mix of an infantry, armour and artillery combined arms team was the first use of a flanking seaborne assault by the Provisional Government forces. The landing itself took place on 24 July, and met no opposition to its occupation of Westport. Indeed, the welcome by local residents was reported to be enthusiastic. As there was no need for artillery support, the 18-pdr was recalled for use elsewhere.

On 24 July, following the seizure of Waterford, Prout began his advance on Carrick-on-Suir; the force consisted of 600 men, and armament included field pieces, trench mortars, and machine-guns.[18] This was the force that would ultimately engage in the fighting over a week later, on 2 August.

On 26 July, in South Tipperary, a National Army force surprised the Republican garrison in Golden (between Tipperary and Cashel) and took the town. A Republican attempt to retake the town three days later, on 29 July, proved to be unsuccessful. The Republican attack force consisted of some 400 men, supported by at least one Lancia armoured troop carrier, and split into three columns. The government forces broke up the attack with the aid of an 18-pdr, very likely the same 18-pdr under Galvin's control that supported the attack on Limerick the previous week.

On Friday, 28 July, government forces attacking Kilmallock again were accompanied by artillery. The field gun was brought to Dromin on a lorry, and then lowered on sloping planks to the road. Two horses then were hitched to the gun; reportedly, one was an unwilling cart horse, while the other, by some combination of circumstances, was an old Royal Field Artillery horse that took eagerly to the traces and its familiar task. The gun was taken up a muddy lane and over hedges and ditches to a previously chosen position, unlimbered, and brought into battery.[19] From the hill at Dromin and from a range of about three miles,[20] it opened fire against enemy trenches at Kilmallock Hill outside the town itself, but the shells overshot their targets. Shrapnel fire was then directed at a house suspected of being occupied by Republicans, but results were inconclusive. Fire was later brought to bear on a quarry on Kilmallock Hill that housed an Irregular post; firing this time was particularly accurate, two out of three shells dropping into the quarry and the third

15. *Calton Younger, page 381.*
16. *Michael Hopkinson, Green Against Green, The Irish Civil War, page 150.*
17. *Eion Neeson, page 147. No details as to the number, calibre, or type of mortar are provided.*
18. *Hopkinson, page 168. Although Hopkinson makes references to "field pieces" (plural), it does not seem likely that there was more than one 18-pdr available to Prout. What is interesting, however, is the specific reference to trench mortars. Neither number, type, provenance or use of these mortars are mentioned, although these may be the same mortars referenced earlier during the attack on Limerick. The only other references to mortars discovered during the author's research were to a Republican mortar nicknamed "Lizzie" used at Bruree on 2 August (Neeson, page 205), and a rudimentary trench mortar used at Castlemaine in early January.*
19. *The Irish Times, 5 August, 1922.*
20. *This was well within the range of the 18-pdr, but far greater than the ranges normally experienced in most of the fighting, which typically was in urban areas. The artillery tended to fire at more or less close or even point-blank range.*

bursting just over it, achieving the desired effect. The National Army troops were able to continue their advance.

On 30 July, National Army forces attacked Bruree, approaching from two directions. The artillery that had been used to decide the battle in Limerick some ten days earlier was brought in from Bruff to the north-east. The Republicans were forced to withdraw, although three days later they returned, supported by two improvised armoured cars. During the 30 July attack, Galvin's field gun was used to fire on Republican rear guard forces that had been holding a bridge. If one discounts the serious fact that men on each side were shooting at each other, usually with the intent to wound or kill, a bit of *opera buffa* played itself out with the 18-pdr at Bruree. According to Eion Neeson, the gun crew had dug a hole so that the trail of the gun could be properly anchored and a shell had been rammed into the breech, ready for firing, when a herd of cattle wandered across the line of fire only a few feet from the gun's muzzle, obliging the gunners to delay firing until they could drive the cows away.[21]

On 1 August, Prout's advance that had begun from Waterford on 24 July had finally drawn close to Carrick-on-Suir; at Piltown, some four miles east of Carrick when it made contact with Republican forces that, coming from Fiddown Bridge a mile or so to the south, had taken positions in the woods near the river to his rear. Small arms and machinegun fire was inconclusive, but the following day, 2 August, the 18-pdr was brought up and shelled the woods with shrapnel, causing the Republicans to withdraw. Apparently Prout's force then swung north, as the 18-pdr fired a couple of shells into the castle at Kilcash to flush out any possible defenders.

On 2 August, a seaborne attack conceived by General Emmet Dalton and led by Brigadier Paddy O'Daly was mounted against Fenit on Tralee Bay, supported by the armoured car *Ex-Mutineer* as well as by an 18-pdr named *The Rose of Tralee*. The motorized prime mover for the 18-pdr had been damaged during the landing, making it necessary to make other arrangements for towing the field piece. The National Army soldiers impressed a horse and a mare named Dolly, as well as tackle and chains, from a local farmer named Timothy Egan, hitched the horses to the gun, and moved off under protection of the armoured car. *The Rose of Tralee* seems not to have been engaged in action during the battle for Tralee itself, although reportedly it was brought into action a few days later during the advance on Castleisland, firing about half a dozen shrapnel shells, and again later in the month at Droum, near the Pap Mountains beyond Killarney.

During the first few days of August, following the seaborne attack against Fenit on 2 August, another force under Brigadier James Slattery supported by the now familiar combination of armoured cars and artillery quickly took Adare in County Limerick (where resistance was quickly overcome with the assistance of the 18-pdr), seized Rathkeale from the Republican defenders, continued its advance southward to Newcastle, and eventually linked up with the Fenit force. Presumably, any Republican opposition along the way was dealt with, at least in part, by the accompanying National Army artillery.

The success of the seaborne landings at Westport and Fenit led Emmet Dalton to plan and execute a much more ambitious operation to seize Cork, employing much the same formula. On 7 August the passenger ship *Arvonia* was commandeered and set sail from Dublin to Cork. Another ship, the *Lady Wicklow*, also was pressed into service for the expedition. Between them, the ships carried two Peerless armoured cars and a Rolls-Royce armoured car, as well as several Lancia armoured troop carriers; it is not entirely clear how many 18-pdrs were carried aboard the ships, but photographs suggest that there were at least two of the field guns[22], either both aboard the *Arvonia*, or possibly one on *Arvonia* and one on *Lady Wicklow*. [23] The attacking force landed early

21. *Neeson, pages 204-205.*

22. *Neeson, page 163, states that "An armoured car and an eighteen pound gun were part of the equipment carried," while photographs in Tim Pat Coogan and George Morrison's The Irish Civil War suggest a somewhat different story with respect to the number of 18-pdrs. A photograph on pages 212-213 shows a Mk I 18-pdr aboard the Arvonia with the inscriptions "Four Courts" and "Drogheda" on an upper gun shield, as well as a photo on page 224 of a Mk II 18-pdr that clearly is not the same gun pictured on pages 212-213 purportedly being unloaded from the Arvonia.*

23. *Calton Younger, page 412, states that an armoured car and an 18-pounder were put ashore from Lady Wicklow.*

An 18-pdr aboard the Arvonia, *bound for the attack on Cork City on 8 August 1922; the photograph likely was taken on 7 August. The names on the upper shield identify this as the first gun in action against the Four Courts in Dublin in the closing days of June, as well as the gun involved in the shelling of Millmount, Drogheda, on 4 July. (NPA HOG 30, Courtesy of the National Library of Ireland)*

on the morning of 8 August, supported by fire from at least one gun. Following some three days of fighting, the Republican forces dispersed towards Macroom, partly in order to avoid confrontation with the government artillery; the Republicans had learned well the lessons from earlier battles in urban areas and had decided to husband their strength rather than subject themselves to an attack against which they had no remedy.

National Army troops also advanced towards Douglas, between Cork and Cobh. The oft-repeated combined arms formula was again in evidence, with the infantry column being joined and supported by the armoured car *Manager* as well as by a truck carrying an 18-pdr; Douglas was soon cleared of any Republican defenders.

On 18 August a large motorized column left Killarney for Rathmore, intending to carry out a reconnaissance of the East Kerry area. Upon reaching Barraduff, the column split in two, with one part advancing towards Rathmore while the other proceeded towards Headford. At Droum Bridge, the Republicans dug a trench, and when the government column reached the point, took it

The funeral of Michael Collins, 28 August, 1922. Michael Collins' casket is borne on top of one of the 18-pdr field pieces so effectively used by the National Army during the Civil War. At the time of Collins' death, there were eight 18-pdrs in the National Army inventory. (Broe Collection)

under small arms fire, wounding the leader, Colonel James McGuinness. Upon hearing the gunfire, the other government column came to assist, accompanied by an armoured car and by the 18-pdr, which lobbed a few shells against the Republicans, who soon withdrew.

On 25 August a large column under Fionan Lynch left Killarney for Kenmare to strengthen the garrison there. As the column advanced along the Killarney-Kilgarvan road near Glenfesk, it came under insurgent fire. The column included an 18-pdr (probably the *Rose of Tralee*), but due to the terrain, which was marked by steep cliffs, the field gun was useless against the Republicans, who had chosen their ground carefully. The government force was obliged to withdraw – one of the few instances in the war in which the National Army's artillery was not able to decisively influence the outcome of the engagement.

On 10 September a National Army column was ambushed between Killarney and Rathmore, killing a number of soldiers. Apparently the column was accompanied by a field piece, as the Republican attackers were driven off after five shells had been fired.

Perhaps the last action involving one of the 18-pdrs occurred in late October (sometime around or after 27 October) at Ballyheige in County Kerry. National Army troops had surrounded a group of Republicans who put up a stout resistance; the government troops called for artillery support, and an 18-pdr from Tralee (again, in all likelihood the overworked *Rose of Tralee*), which predictably resolved the issue in favor of the besieging force.

Throughout the conventional phase of the fighting during the Civil War (beginning on 28 June 1922 in Dublin, until roughly September of the same year), there was no formal artillery establishment in the National Army. Much as had been the case with the National Army's armoured cars, which were parceled out to individual commands or commanders, the same was true of the limited artillery assets, ranging from an initial issue of two 18-pdrs for the attack on the Four Courts, increased to four guns during the fighting in Dublin, to eight pieces by mid-August. The guns were allocated to the commands[24] and seem to have been employed effectively in most, if not all, instances.

A closer view of the 18-pdr bearing the coffin of Michael Collins. All of the brass components and fittings on the gun have been polished to a mirror finish for this sad but historic occasion. (Broe Collection)

The actions described allow at least a tentative identification of most of the guns involved in the fighting. It is clear that initially there were two guns in Dublin on 28 June, and by 2 July there were four. The pace of delivery of the remaining five guns is unclear, but there is no doubt that as soon as they were received from the British, the National Army committed them to the support of forces in the field. Once the Dublin fighting was concluded, the first gun to make its appearance – or rather, reappearance – was the gun under Johnnie Doyle at Millmount on 4 July; this was the same gun that had begun shelling the Four Courts from Winetavern Street on 27 June. The second gun that can be identified was the one under Sean MacEion's command, in Boyle on 5 July, followed by action in Collooney on 13 July. The third gun that can be tracked is the one that accompanied the column under Commandant Bishop that moved to Blessington, through Carlow and Wicklow on 6 and 7 July. The fourth gun was the one used in the attack on Waterford beginning on 19 July, under Prout's command, which on 2 August was used to support the attack on Carrick-on-Suir. The fifth gun, under the command of Dinny Galvin, supported the attack on Limerick on 19 July, followed by actions in Golden on 26 July and Kilmallock on 28 July, and most likely was the gun in action at

24. The number of National Army commands changed rather quickly: on 5 July, there were three District Commands (Eastern, Western, and Southern), and by 28 July the number had risen to six (Eastern, South Western, Curragh, 2nd Southern, Western, and 1st Northern). By the end of August, there were eight commands (Eastern, Cork, Kerry, South Western, Western, Northern, Waterford, and 3rd Southern).

Bruree on 30 July. Although it is conceivable that this gun was also the gun described in action at Adare and Rathkeale on 2 August, and possibly at Dromin on 4 August, as well as the 10 September actions at Killarney and Rathmore, it is equally possible that these actions were attributable to another gun (in fact, the gun at Adare was noted as being under the command of Brigadier James Slattery, not under Dinny Galvin). The sixth gun, the *The Rose of Tralee*, supported the landing at Fenit on 2 August, and continued to operate in Kerry, probably at least until late October. This gun may have been the same gun that, having been assigned to support the 22 July landing at Westport but found to be superfluous, was then included in the Fenit attack force. Two guns were allocated to the landing at Cork City on 8 August; one of those guns can be confirmed as the original Winetavern Street gun, and the other (the seventh) may also have been one of the guns that saw action in Dublin, possibly the companion gun at Winetavern Street on 30 June. Allowing for the possibility that one or two of the guns in the above reckoning may have been double-counted, a total of at least six of the nine guns in the inventory by the end of 1922 guns can tentatively be accounted for in various actions.

Although for historical purposes it would be highly interesting to be able to determine the precise number of artillery rounds fired during the Civil War, arriving at such a figure is manifestly impossible. A very rough order of magnitude, however, can be guessed by beginning with some known figures and extrapolating others. The sheet recording the number of rounds fired by the first of the Army's guns (the Winetavern Street gun) reflects a total of 440 rounds fired between 27 June and 8 August 1922. It is a given that more rounds were fired against the Four Courts than against any other single target during the war, so it would seem safe to assume that no other gun approached the number of rounds as fired from that gun. The devastation caused by artillery fire against buildings on and near The Block in Dublin may have accounted for another 50-100 rounds maximum. A figure of 36 rounds is reported to have been fired at Waterford, although this may be an incomplete figure. Other known figures include 33 rounds fired against Strand Barracks in Limerick; a rough guess for the total number of rounds fired in Limerick would be about 85 rounds. At least six rounds were fired by The *Rose of Tralee* at Castleisland. Added to these actions were a number of other instances in which artillery was used (Golden, Kilmallock, Bruree, Carrick-on-Suir, Killarney, Cork and a few others), which might bring the total up to something on the order of 725 to 750 rounds. Aside from the shelling of the Four Courts and the relatively major actions at Waterford and Limerick described above, there simply were not that many targets that required sustained shelling. Well over half of the estimated total of rounds were fired during the siege of the Four Courts and the subsequent fighting in Dublin; during the two month period of July and August, perhaps another 300 rounds were fired, amounting to an average of only five rounds a day. The small number of shells expended again points to the overall effectiveness of the National Army artillery, as well as the lack of any meaningful defense against it by the Republican side.

Belatedly, in October Liam Lynch, the Republican Chief of Staff, put forth a proposal that artillery for the Republican forces be purchased from abroad. Feelers were put out in both the United States and Germany, and apparently some thought was also given to the possibility of purchasing guns in England, as implausible as that may have been in view of British support for the Provisional Government. By January 1923 a rather bizarre scheme was developed to obtain sixteen mountain guns and four heavy guns from Germany, to be landed in Munster by submarine or by air in order to avoid naval patrols.[25] The project, however, was compromised by the capture of correspondence that alerted both the British and Free State governments to the plan; this, coupled with financial and practical difficulties, put an end to the whole affair.

25. Details as to the type of guns being considered are unavailable. The most likely candidate for the mountain gun would have been the 7.5 cm (75mm) Erhardt GebirgsKanone M.11 produced in 1910 and supplied to Norway. However, the availability of such guns is open to question, as was the ability to deliver them to Ireland via submarine or aircraft. Although the Erhardt gun could be broken down into six loads for transport, its overall weight would make delivery by air somewhat problematic. Delivery of heavy guns by air or submarine would not appear to have been feasible. Also, in 1922-1923, Germany was forbidden by treaty to have submarines in its navy, so just whose submarines would have delivered the guns remains a mystery.

Names of guns

The practice of bestowing names to large items of military equipment such as tanks, armoured cars, and artillery is widely accepted, although not always officially sanctioned, in most armies. During the Irish Civil War most armoured cars on both sides bore names, and similarly, at least some of the National Army artillery pieces were embellished with inscriptions by their crews. Of the nine 18-pdrs in the National Army inventory at the time, at least three names have been observed or reported. The first gun to be used in the assault on the Four Courts and later on the attack at Millmount, Drogheda appears in a photograph taken aboard the *Arvonia* en route to the attack against Cork City; the inscriptions *Four Courts* and *Drogheda* are painted on the upper portion of the shield, commemorating the gun's actions at those locations. Another 18-pdr, probably photographed in Dublin during the siege of the Four Courts, bears the name *Hammond Lane No. 4*; this author has been unable to determine the significance of the name, although Hammond Lane is the street immediately to the west of the Four Courts. The third gun to be associated with a name was the18-pdr that took part in the landing at Fenit and subsequent attack on Tralee; not surprisingly, it was christened the *Rose of Tralee*. It is likely that at least a few of the other nine guns in the inventory were adorned with names at some time or other during the campaigning, but if so, they remain unrecorded.

The use of artillery by the Republican side was almost nonexistent. They did reportedly have one mortar nicknamed "Lizzie" (not specifically identified by type, but possibly a Stokes trench mortar) and used during the fighting in Bruree on 2 August, but when, where, and how it was obtained are all open to question. There also is some scant indication that as late as January 1923, well after the conventional phase of the fighting was over, their armourers managed to develop a rudimentary trench mortar.[26] On 23 January, an anti-Treaty column attacked the military garrison at Castlemaine, during which the effectiveness of the mortar was tested. The experiment was largely unsuccessful, due to a combination of inaccurate targeting, defective ammunition, and ammunition failing to explode because of soft, boggy ground.

If the Republicans could be said to have had one notable success of sorts with respect to artillery it would have been the 14 August surprise attack on the National Army forces in Dundalk in which a variety of equipment including rifles, machineguns, ammunition, and grenades were captured, but the outstanding prize was the capture of an 18-pdr. The seizure of the field gun was somewhat of a transient victory, however, as lack of trained personnel to crew it, as well as problematic ammunition availability, induced the Irregular forces to destroy the gun rather than attempt to use it, or risk it falling once again into the hands of government forces that would again be able to employ it against them. The loss of one of the 18-pdrs by this time would have been somewhat less critical than it would have been only a few weeks earlier. The conventional phase of the fighting had all but ended, with the Republicans soon to revert to guerrilla-style warfare for the most part. Added to that was the fact that by mid-August the British had turned over a total of eight of the 18-pdr guns, essentially doubling the number that had been employed by Provisional Government forces during July and early August.

Throughout the period of the Civil War, the fact that the National Army was equipped with artillery proved to be a, if not the, decisive factor in operations against the anti-Treaty forces. In terms of equipment, the National Army held a decided advantage, fielding much heavier weaponry than the Republicans, especially with respect to armoured cars and artillery. Although the Republicans managed to capture a few armoured cars and improvise others which enabled them to at least compete with the National Army on several occasions, there was no corresponding counterbalance to the artillery in the National Army's inventory – other than possibly two mortars,

26. *Tom Doyle, The Civil War in Kerry, page 251.*

The Artillery Corps Logo

the Republicans simply had none. Once the National Army guns and gunners were brought on the scene, the Republicans had the choice of withdrawing to other positions or to hunker down in the most solidly constructed buildings they could find, which inevitably became targets, and which just as inevitably eventually were reduced to rubble by the guns. There was, of course, no counterbattery fire to worry the government gunners; although Republican snipers occasionally eliminated individual gunners, they also had to face the prospect of gunners who thought it to be great sport to engage the snipers with their 18-pdr guns, as was reported on at least two occasions.

The Artillery Barracks in Kildare, originally established by the British in 1901 as a Royal Artillery barracks, was taken over by the Artillery Corps on 20 March 1925. In 1952 it was renamed in honor of Sergeant Magee, but in 1998 was vacated by the Army; it had served as home of the Artillery School, the 3rd Battery (PDF), the 1st Air Defence Regiment, and the 6th Field Artillery Regiment (FCA). Plans for its redevelopment as housing units had not materialized as of 2012.

Gunner Magee and the Magee Trophy

The story of Gunner James Magee and his companion, Gunner Casey, is the type of inspirational feat that is the stuff of legend, and holds a special place in the history and memory of the Irish Artillery Corps.

In 1798 the United Irishmen, with French help, attempted to gain their independence from England. A small French expeditionary force landed in Ireland, and on 27 August a combined Irish and French force numbering about 2000 routed a British force of about 6000 at Castlebar. This victory, however, proved to be fleeting, as less than two weeks later the Irish and French again faced off against the British at Ballinamuck in County Longford. It was during this battle that an epic episode took place involving Gunner James Magee and Gunner Casey who somehow or other had managed to acquire two French 6 pounders, which represented the sum total of the Irish insurgent artillery. On 8 September 1798 Magee and Casey and their two guns and crews faced the British forces under Cornwallis at Ballinamuck. Although Franco/Irish hopes had been buoyed by the earlier victory at Castlebar, when the small French/Irish force met the opposing British forces at Ballinamuck, there was little hope that they could prevail against the might of the arrayed English forces. Once battle had been joined, Magee and Casey brought their guns into action, trading shots with the English gunners. A round from one of the English guns disabled Casey's 6 pounder, leaving Magee's gun as the sole piece to oppose the English. Magee's gun continued to wreak havoc on the opposing English forces, but ammunition stocks ran low; the desperate but ingenious gunners broke up pots and kettles into shards and mixed these with grape and canister and thus managed to stave off a tentative English cavalry charge. While the gun crew were in the process of reloading, an English cannonball struck and shattered the stock of one of its wheels. Three of the gun's crew rose to the challenge by hoisting the now loaded gun on their shoulders. Gunner Magee touched the match to the tinder for the last time, the gun fired, and the three crew were killed by the recoil. Magee himself was captured, court-martialed in the field, and condemned to execution by hanging.

The Field Artillery Trophy commemorates the heroic stand of the Irish gunners at Ballinamuck, a tribute to the ability of the gunners to keep the guns firing at all costs. It is a testament and a reminder to all of the members of the Artillery Corps to the dedication and sacrifice of their predecessors. The trophy itself is a bronze replica of the French six pounder, complete with the damaged right wheel, used by Sergeant Magee at Ballinamuck. It resides in the Artillery School offices in the Defence Forces Training Centre.

Field Artillery

Irish gunners crewing an 18-pdr on a Mk V carriage, distinguished by its split trails. The wood spoke wheels and the Irish 1928 pattern helmet made by Vickers that was modeled on the German M1916 "coal scuttle" helmet date this photograph sometime prior to 1939, when the 18-pdrs were upgraded with Martin Parry carriage conversions with pneumatic tires. (Military Archives via Lt Col Edmond Fogarty)

The Artillery Corps (*An Cór Airtléire*) was formally established on 23 March 1923, when the nine available guns were grouped together at Islandbridge Barracks (now Clancy Barracks), Dublin. The Corps had a strength of eleven officers, sixty-two Other Ranks, and ten horses, and the nine guns consisted of five Mk I and four Mk II 18-pdrs.

On 10 July 1923, the embryonic Artillery Corps took the first step to develop organisationally along traditional artillery lines, grouping four of the 18-pdr guns (Mk II) into the 1st Field Battery. The battery, stationed at McKee Barracks, had a strength of five officers, eighteen NCOs, and 93 gunners. The 2nd Field Battery, with four Mk I 18 pdrs, was formed on 5 January 1925; shortly afterwards, on 20 March 1925, the Corps transferred from McKee Barracks in Dublin to the Artillery (Magee) Barracks in Kildare, which became its permanent home. The 2nd Battery is credited with the first shoot in the Glen of Imaal on 10 June 1925, although other sources claim that the first firing practices began either on 1 September, or on 11 September, 1925, by the 1st Battery. During the course of 1925, four British QF 4.5 inch howitzers were added to the inventory, equipping the newly formed 3rd Field Battery. In April 1926 the Director of Training recommended that the guns be painted grey instead of the green that had characterized them until that time. In January 1928 two Mk IV 18-pdrs were added, followed by another eight Mk V 18-pdrs in December 1930. Another four-gun battery of 4.5 inch howitzers was acquired in October 1929. There also is some suggestion that six 18-pdrs and six 4.5 inch howitzers that were part of the British coast artillery defence armament in 1938 were turned over to or bought by Ireland when the coast defence forts were relinquished by the British at that time.

From its earliest days, the Artillery Corps, as small as it was, strove for excellence. As early as 1928 the artillery was considered to have the strictest discipline in the Army. There was no

An upgraded 18-pdr on the Martin Parry Mk II P carriage. This gun was one of the early Mk II 18-pdrs, very possibly one of the original nine guns in the inventory in 1923. The single pole trail has been retained, although the tires are now pneumatic. Vehicles in the background include a Beaverette light armoured car and three Morris-Commercial C8 field artillery tractors.(Military Archives)

A restored 4.5 inch howitzer at McKee Barracks, Dublin, in October 2009. The first howitzers of this type were delivered to the Irish Army in 1925, forming the third battery of the young Artillery Corps. (Sean Reilly, via Ian Stewart)

Perhaps not surprisingly, a number of officers and NCOs in the fledgling Artillery Corps had previous artillery experience with the British and Canadian armies during World War I. Among those who had served with artillery units during the Great War were James McLoughlin, Harry Frowd St. George Caulfield, Garrett Brenan, Bertie Thomson, Sergeant Major W. Downey, and William Bonar. McLoughlin was from Enfield, Co. Meath and enlisted with the Royal Garrison Artillery in November 1915 and served with the 103rd Siege Battery from 1916-1918 in France. He was wounded in March 1917. After the war he served in the Sudan and the Coastal Defences at Cork as Sergeant and enlisted in the National Army in August 1922. He was posted to the Artillery Corps in October 1924 as Officer in Command of No. 1 Battery and served as Director of Artillery from 1949-1955. St. George Caulfield was commissioned into the Royal Field Artillery in 1900, serving until 1907. He was recalled into service in February 1915 as an Acting Captain and served in the UK with 3A Reserve Brigade, Royal Field Artillery as an Acting Major. Brenan was born in Co. Kilkenny and served with the Antrim Royal Garrison Artillery (Special Reserve) as an officer from 1906, and promoted to Captain in May 1909. He was posted to France in 1916 and promoted to Acting Major in command of a battery in December 1916, serving with the 79th Siege Battery, 67th Brigade. He was awarded the Croix de Guerre on 23 July 1918 and was also twice mentioned in dispatches in 1918. He was appointed Instructional Officer with the Artillery Corps in 1926. Thomson, born in Chatham, England, served with the Royal Canadian Artillery during the war as Regimental Sergeant Major, and in October 1916 was granted a temporary commission in the Canadian Field Artillery. He later served as an Instructional Officer with the rank of Captain prior to transfer to the Artillery Corps, and served as commander of the 1st Field Artillery Brigade in the 1930s. Downey served with "K" Battery of the Royal Field Artillery during the war; he was Company Sergeant Major with the 3rd Field Battery in 1937 and served as Regimental Sergeant Major from 1943 to 1946. Bonar was born in Killygordon, Co. Donegal and enlisted in the British Army in 1918 at the age of 16, serving in Mesopotamia during the Arab rebellion of 1920-21 with 2nd Battery, 13th Brigade, Royal Field Artillery. Following enlistment in the National Army in 1924 he was posted to No. 1 Battery at Kildare. He served with the 4th Field Artillery Regiment in Mullingar during The Emergency, and retired in 1947 with the rank of Company Sergeant. (Information courtesy of Mark McLoughlin)

Officers of the Artillery Corps, Kildare, Spring 1931. Front Row: Capt Charles Trodden (Artillery School), Capt. James McLoughlin (OC 1st Field Battery), Comdt. Patrick Maher (OC 1st Field Artillery Brigade), Major Patrick A. Mulcahy (Officer-in-Command, Artillery Corps), Capt Richard J. Callanan (Quartermaster), Capt. Bertie S.C. Thomson (OC 2nd Field Battery), Capt. M.O. Carroll; Second Row: 2/Lt P. J. Halley, Lt. D.J. Farrell, Lt. J.J. Dolan, Capt. Denis J. Cody (OC 3rd Field [Howitzer] Battery), Lt. Thomas J. Lambert, 2/Lt. J.L. O'Brien, Lt. W. P. Donnelly, Lt. M.P. McCarthy; Third Row: 2/Lt. Arthur Dalton, Lt. Peter P. Farrell, Lt. C.F. Byrne, Lt. P. Doyle, Lt. M. Buckley, 2/Lt. P. Curran, 2/Lt. W.A. Moloney, 2/Lt. K. Curran (Mark McLoughlin)

tolerance for anything other than the highest standards of training, drill, and dress. Brass on guns and on uniforms was constantly burnished, and leather belts, boots, and bandoliers were polished to perfection. The drivers, as the horse soldiers were known, spent untold hours dedicated to the care, feeding, and grooming of the horses, and, in addition to caring for their own personal leather kit, had to maintain all of the accoutrements for the horses as well. All of this spit and polish and attention to detail and procedure never failed to impress, and the Corps was much in demand for ceremonial and festive occasions such as St. Patrick's Day and Easter parades.

The pomp and majesty of the horse soldiers reached its ultimate expression in the creation of a very colourful mounted escort for ceremonial duties. The Army began planning to form a mounted escort during 1931; the unit first appeared in public in 1932 escorting the Papal Nuncio, Cardinal Lauri during the Eucharistic Congress held in Dublin 21-26 June. The unit was formed on an *ad hoc* basis, mainly with personnel from the Artillery Corps, which comprised the largest body of horse soldiers in the Army at the time. The mounted escort unit bore the official title of *An Marc Sluagh* in Gaelic, but was more colloquially referred to as the Blue Hussars in reference to their colourful uniforms consisting of a rich sapphire blue tunic (officially, alizame sapphire) with yellow piping and frogging, white breeches with a yellow stripe, and Hessian-style black boots. The uniform was completed with a black sealskin

An impeccably kitted out Gunner Gus Jones, standing next to a 25-pdr. Soldiers who strove to be as smart and sharp as possible were referred to as "jildy" – the word jildy had come from India, via British soldiers, and referred to a particularly well turned out soldier. Gnr Jones is wearing the British "Tommy" helmet, and his sharply creased trousers may have had lead weights in the bottom to keep them hanging properly. The Artillery Corps has long maintained a reputation for high dress standards. (Mullingar Barracks Museum via Colin Stone)

A Blue Hussar trumpeter. The uniform is the same as worn by a regular trooper, but also incorporates additional adornment in the form of prominent shoulder epaulettes and a braided lanyard that crosses the chest. (Military Archives)

The colourful and ornate Blue Hussar tunic and sealskin busby. This uniform, designed to be suitably impressive for ceremonial occasions, was tastefully elegant but avoided being gaudy. (Colin Stone)

busby with saffron yellow embellishments. Further adding to the impressive appearance were the lances carried by the mounted soldiers. The escort, consisting of about eighty horses, was not a permanent standing unit, but was brought together for special ceremonial occasions. The escort consisted of an advance guard of two, a single connecting file, two flanking riders, and two troops of thirty. From 1939 to 1945 its activities were completely curtailed, and were briefly revived in 1945. In 1949 it was definitively disbanded, due to lack of a sufficient pool of trained horse soldiers to draw upon, and the escort duties were handed over to the Cavalry Corps which, totally lacking horses itself, rode on motorcycles.

The creation of a showpiece unit did not detract from modest progress in expanding the Artillery Corps itself. By 1933 the Army had increased the inventory of 18-pdr field guns, bringing the total of that type of gun on hand to 25.[1] That same year, four Ordnance QF 3.7 inch pack howitzers were acquired from Britain and were formed into the 1st Light Battery, which one source says also included the limited number of light mortars available at the time, although neither the number nor type of mortars is cited – these may have been a few Stokes mortars remaining from the Civil War, as the earliest credible reference to acquisition of mortars by the Army is 1936, when thirty-one 81mm mortars were obtained in the March-April timeframe.

Defence Force Regulation 49 of 22 October 1934 established "The Tactical and Territorial Organisation of the Defence Forces." It authorized a Permanent Force with four batteries organised as the 1st Field Artillery Brigade, consisting of the 1st, 3rd, and 4th Field Batteries of the Permanent Force, and the 2nd Field Battery in reserve. Each battery had four guns; the 3rd Battery was equipped with the 4.5 inch howitzers, while the other three batteries were equipped with 18-pdr pieces. The Volunteer Force, formed in March 1934 and consisting of nine territorial regiments provided an additional twenty-four field batteries nationwide (the 5th through the 28th), and six light batteries (the 2nd through the 7th). A camp diary for Coolmoney Camp covering the period from May 1936 to September 1941[2] contains numerous references to training for the 5th through the 16th Field Batteries (Volunteer Force), as well as to the 2nd through 4th Light Batteries (Volunteer

1. *It can be fairly conclusively established that by 1930 there were nineteen 18-pdrs in the inventory (nine on hand in 1923, two delivered in 1928, and eight in 1930), but how many guns were delivered by 1933 is somewhat elusive; extrapolating information from several sources leads to the figure of twenty-five guns by 1933, with six of the guns delivered in 1933 bringing the total to 25, although at least one other source indicates that only three guns were delivered in 1933.*
2. *Camp Diary, Coolmoney Camp, 18th May 1936 to 5th December 1941.*

A 3.7 inch mountain howitzer as restored, date unknown. This is one of only four such guns acquired by Ireland in 1933, and retained in the inventory until the 1970s despite their age. (MA, Artillery Box)

Force). The 2nd Field Artillery Brigade, organised in 1939, was a reserve formation, and incorporated the 5th through the 8th Field Batteries (Volunteer Force); the 5th through 7th batteries were each equipped with four 18-pdrs, while the 8th appears to have been a 4.5 inch battery. Manifestly, however, there were not nearly enough guns in the inventory to equip a total of twenty-eight field batteries (Regular and Volunteer force combined) plus the seven light batteries. An establishment of four guns per battery would require 112 guns, plus twenty-eight for the light batteries, or a cumulative total of 140 guns. In reality, the number of guns of all types in the Irish Army artillery park fell short of that number by over a hundred guns. The best estimate is that in 1934 there were no more than twenty-five 18-pdrs of various marks, four 3.7 inch howitzers, and eight 4.5 inch howitzers, so that the total of 37 guns would allow no more than eight full four-gun batteries, plus the Light Battery with its 3.7 inch howitzers, to be fielded. The four batteries of the 1st Field Artillery Brigade would account for sixteen guns and the four batteries of the 2nd Field Artillery Brigade for another sixteen guns; the independent Light Battery was equipped with the four 3.7 inch howitzers. The numbers would thus suggest that there were enough guns for six full 18-pdr batteries and two full 4.5 inch batteries, plus the 3.7 inch Light Battery. With the exception of the 5th through 8th batteries in the 2nd (Reserve) Field Artillery Brigade, the Volunteer batteries existed organisationally and had a manning complement but were privy of equipment. Records from 1936 through 1941 show the 5th through 16th Field Batteries and the 2nd through 4th light batteries engaged in training at Coolmoney Camp in the Glen of Imaal, often alongside regular units. Eventually, each Volunteer battery had something on the order of three officers, up to ten NCOs, and some fifty Volunteers trained. During these camps, which usually lasted about ten days, the Volunteer Forces batteries managed to fire hundreds of rounds each. The light batteries in actuality were mortar batteries, equipped with 81mm Brandt mortars which may actually have been on issue to the units. The Coolmoney Camp diary does not reflect any training by the 17th through the 28th field batteries, nor for the 5th through the 7th light batteries, suggesting the possibility that these batteries were more notional rather than actual. In 1943, when the regimental structure was adopted and the active artillery organisation expanded tremendously, the Volunteer Force batteries immediately were able to provide a pool of officers and gunners who had been trained alongside their Regular counterparts, rendering the expansion a much less painful and confusing process than might otherwise have been the case.

Coolmoney House and camp are the stuff of legend and a bit of a shrine for the Irish Artillery Corps until it was razed in the late 1990s, despite a preservation order that had been issued; the order for its destruction is blamed on bureaucratic confusion, as no one seems to know just who authorized tearing it down. The photograph dates back to the time when the building was occupied by the British, as is evidenced by the barely recognizable Union Jack flying from flagstaff. Coolmoney House was reputed to be haunted. (Colin Stone)

18-pdrs, Glen of Imaal, 1960. Although two of the gunners are holding rounds in their hands, the customary ammunition limber is not in evidence. The 18-pdr was the primary field piece equipping Irish artillery batteries during The Emergency. (62 Reserve Artillery Regiment, via Capt Ian Stewart)

A rather large pile of expended 25-pdr brass ammunition casings. Some empty cartridge casings have been hammered into the ground as stakes to prevent the piles of stacked casings from rolling. Since the end of the Civil War in 1922, all artillery ammunition expended in Ireland has been for training purposes, except for a modest amount of anti-aircraft ammunition fired during The Emergency, as well as a sizeable amount of mortar ammunition fired in support of Irish peacekeeping contingents abroad. (Bryan Ely)

THE RAIKES RANGE

During the formative years of the Irish Artillery Corps – less than ten years after its inception – attention was already being given to training that was innovative, realistic and effective, as well as being simple and economical to operate. To this end, the Artillery Corps developed a rather sophisticated terrain model known as the Raikes Range used to develop the skills of artillery observers. The Raikes Range was a very early type of simulator. It was a model made to scale on a raised platform, laid on a wooden lattice shaped in the form of the terrain it represented, including hills, valleys and other prominent terrain features. It was usually of hessian which was coloured as the terrain dictated, with the outlines of fields embroidered on it, and complete with model houses, roads, rivers, bridges, villages, etc. A typical Raikes Range measured 8-10 meters by 8-10 meters, and represented an area of 8-10 kilometers by 8-10 kilometers. Observers were positioned on raised seating 1-2 meters away from the edge of the panorama. Underneath the raised terrain model there was a command post manned by Battery CP staff. Fire Request Orders were initiated as they normally would be in the field, and were replied to by the CP staff using normal Voice Procedure protocols. They then calculated the location of the fall of shot using a direction orientation board and a lateral and horizontal grid system. The exploding projectile was simulated by blowing a puff of cigarette smoke through the hessian from underneath at the calculated location using a piece of very fine copper tubing, enabling the observer to see where the fall of shot would have occurred. A fire mission was indicated by smoke from a tube with six outlets to indicate a battery fire mission. The first Raikes Range was installed at Kildare in April 1931; similar ranges were installed at various other artillery barracks and Reserve unit locations, and continued in use at several locations until at least the late 1980s; the range at Magee Barracks remained in use until the barracks were closed in 1998.

MECHANISATION

In early 1939 the Irish Army tested the first gun that had been modified with a modernized carriage fitted with pneumatic tires. The test was conducted with an 18-pdr upgraded with the Martin Parry[3] carriage by the 1st Battery of the 1st Field Artillery Regiment firing at abandoned houses as targets in the area of what is now Poulaphouca Reservoir in County Wicklow. Despite initial trepidation on the part of the crew who thought that the gun would be uncontrollable on the rubber tires when fired, the test shoot did not result in the kind of catastrophe they had expected. The positive test results led to the entire artillery corps being mechanically drawn by 1940, as the new carriages, which allowed high-speed towing by trucks, were soon retrofitted to the remaining guns (18-pdrs and 4.5 inch howitzers) in the inventory. The conversion kits consisted of two pneumatic steel disc wheels on stub axles carried on a cranked adapter which fitted over the end of the original axletree. Included were the necessary radius rods, brake gear, layer's seat brackets, etc. needed to enable the guns to be safely towed at speeds up to 30 mph. Until the conversions were completed, some of the older guns were carried portee-style in the beds of Ford 1939 and 1940 pattern trucks. This involved manhandling the guns onto the truck bed by sheer muscle power, aided only by a couple of planks that formed a ramp from the ground to the truck bed; there were no power winches, cranes, or other mechanical gear to assist with the loading or unloading. Once mechanization had been completed, an era had passed. Gone forever were the impressive and unforgettable

3. *The Martin Parry company began as the Parry Manufacturing Company (1884-1919) of Indianapolis, Indiana, and was at one time the world's largest manufacturer of carriages. By 1917 their catalog offered 22 different body styles for the Model T Ford chassis, and in 1919 the company merged with the Martin Truck and Body Company of York, Pennsylvania, becoming the Martin Parry Company. In 1930 Martin Parry was bought by General Motors, and the Indianapolis plant was renamed the Chevrolet Body Division. The York plant was largely shuttered but continued to produce small numbers of bodies until the late 1930s, and at the outset of World War II furnished the Allies with conversion kits for artillery pieces. In 1939 Britain awarded a contract to Martin Parry for 2500 conversion kits for 18-pdr guns at a cost of $1,348,000. The conversion kit had been designed and patented by A.P. Buquor; hence, the conversion kit was sometimes referred to as the Buquor conversion kit. The conversion work on the Irish artillery pieces was carried out by the Great Southern Railway Works at Inchicore.*

Two 18-pdr guns of the 6th Field Artillery Regiment at a shoot on 20 July, 1960. Both guns have Martin Parry carriages, but the two guns have different style axle hubs. The gun tube of the gun on the left is at its maximum recoil, while the tube of the gun on the right is in the normal position, with the breechblock open and ready to receive a fresh round. (62 Reserve Artillery Regiment, via Capt Ian Stewart)

sight of a team of six noble beasts straining at the gallop in their harnesses with an 18-pdr field gun trailing, and the sounds of steel-shod hooves and steel-rimmed wooden artillery wheels clattering over Irish cobbles. Progress undoubtedly it was, and inevitable it was as well, but as it was happening there was many a driver whose throat choked up at the thought of the passing of the horse era in the Irish Artillery Corps. It is especially ironic that Ireland, with its long and proud equine tradition, began to retire its artillery draught animals before the war, while most of the Continental European armies still relied on animals to tow their artillery to some degree during the conflict.

The Emergency (1939-1945)

The World War II period, from September 1939 until 1945 was known in Ireland as The Emergency. By April 1940 the field artillery inventory consisted of twenty-five 18-pdr field guns, four 3.7 inch and eight 4.5 inch howitzers. There was also a lone Mk II 2-pdr anti-tank gun. The supply of mortars was limited to a single 60mm mortar and 48 Brandt 81mm mortars.

During the early part of this period, the artillery organisation, which began to develop along battalion lines, was tailored to Irish circumstances. The Southern Command in Cork had three Mobile Columns (in Tralee, Limerick, and Templemore), each of which consisted of a reinforced company with two 81mm mortars and a Field Artillery Section with two 18-pdrs, as well as the 1st Field Artillery Battalion (in the 1st Brigade, Cork), with the 1st, 2nd, and 3rd Field Batteries. Western Command in Athlone, much like Southern Command, also had three Mobile Columns (in Galway, Sligo, and Athlone), but the Western Command columns were stronger in artillery, each with a field battery (the 5th, 6th, and 8th, respectively) with four 18-pdr guns. The General Reserve in the Curragh had the 1st Light Battery with its four 3.7 inch pack howitzers.

Following the Franco-British debacle at Dunkirk in May-June 1940 Germany offered Ireland large quantities of captured British equipment, including 46 field guns, but the offer was declined for a number of reasons. However, in 1941 Britain itself supplied Ireland with additional artillery, including four of the venerable 18-pdrs, four 4.5 inch howitzers and eight ex-French 75mm guns. These guns were enough to equip four batteries, but fell far short of requirements; two years earlier, in 1938, among other items of equipment, projected artillery needs were thirty-two 2-pdrs (although only one was on order), twelve 4.5 inch howitzers, sixty-four 40mm Bofors guns, sixteen medium antiaircraft guns, and sixteen 81mm mortars.

In 1940 the field artillery brigades had been replaced by a field artillery battalion structure, and the Artillery Corps organisation consisted of a Director of Artillery at the GHQ level who controlled the Depot Artillery and School in Kildare, four field artillery battalions, the 3.7 inch howitzer battery in the Curragh, an anti-aircraft battalion, four anti-aircraft/anti-tank batteries, and the coast artillery units. The field artillery battalions, which incorporated newly obtained equipment, were the 1st Artillery Battalion with the 1st, 2nd, and 3rd Field Batteries of the 1st Brigade (Cork), the 2nd Artillery Battalion with the 7th and 10th Field Batteries of the 2nd Brigade (Dublin), the Curragh Command's 3rd Brigade's 3rd Artillery Battalion with the 4th, 9th, and 12th Field Batteries, and the 4th Brigade's 4th Artillery Battalion with the 5th, 6th and 8th Field Batteries at Mulllingar. Each of the four field artillery battalions had three batteries (except for the 2nd, which had only two batteries), each

A photograph taken during The Emergency of an 18-pdr on the Martin Parry carriage during a training exercise. The utility of using branches as camouflage for a gun positioned at an urban street intersection is highly questionable, although the muddy tires of the carriage suggest that the gun previously may have been deployed in a field location where the foliage would have been appropriate, but was not removed when displacing to the current location. The civilian bystanders exhibit varying degrees of interest in the proceedings. (MA, Emergency Box)

A practice drill with the 4.5 inch howitzer at Collins Barracks, Dublin circa 1944. The gun crew, as well as the gun, are in immaculate condition. This type of barracks drill serves to acquaint personnel with proper positions, procedures, and sequence of events for firing a gun. The gunner kneeling behind the trail, with his arm raised, is the gun commander, Sgt Terry Lowen. (Brian Lowen)

with four guns. The anti-aircraft/anti-tank batteries bore numbers matching those of the field artillery battalions, but were largely notional, as in 1940 the Irish Army possessed no more than fourteen assorted air defence weapons and only one 2-pdr anti-tank gun; the anti-aircraft guns equipped the batteries of the anti-aircraft battalion.

From 1941 to 1943 Ireland continued to expand its artillery inventory, enabling it to equip an ever expanding number of batteries and significantly improving its capabilities. In 1941 eight ex-French 75mm guns were received from Britain as well as four 18-pdrs, and four 4.5 inch howitzers, and in 1942 two 60-pdr guns, eight 3.7 inch anti-aircraft guns, and two 6 inch coastal guns for a new fort at the mouth of the Shannon were delivered. By 1943 a further twenty 4.5 inch howitzers, four 60-pdrs, four 3.7 inch and two 40mm anti-aircraft guns, and six 2-pdr anti-tank guns had been supplied by Britain. In 1944 fifty 3-inch mortars were received..

In 1943 the field artillery battalions were replaced by a regimental organisation. The Volunteer Force batteries, almost all of which had existed on paper but which had no equipment, were absorbed into the active establishment or had their numbers assigned to newly constituted active units. The size of the artillery establishment was increased considerably; the four field artillery battalions were replaced by seven field artillery regiments numbered 1st through 7th; each regiment had three subordinate batteries. The designation of the batteries was changed from field batteries to field artillery batteries, and the batteries were numbered 1st through 21st, but there was no pattern as to which numbers were assigned to which regiments, i.e., 4th Field Artillery Regiment had the 8th, 15th, and 20th batteries. In addition to the regimental batteries, there were at least three independent batteries: the 22nd was subordinate to Western Command, and the 41st and 42nd Batteries were based in Dublin. Each regiment consisted, at least in theory, of three field artillery batteries, and one anti-aircraft/anti-tank battery; the AA-AT batteries, however, were not formed due to lack of the necessary equipment (some had only a single 2-pdr anti-tank gun). Authorized manpower for a regiment was 456 personnel. Equipment issue was standardized insofar as possible, given the diverse types of guns available. The field artillery batteries that constituted the basic unit of the Artillery Corps consisted of four pieces. Generally, each regiment had one battery of howitzers and one battery of 18-pdrs; the third battery could be either 18-pdrs, 4.5 inch howitzers, or 75mm guns. The Coast Artillery batteries had no comparable standard establishment given the disparity in the size and fixed equipment of the various forts, although generally speaking each fort had a minimum of two 6 inch or 9.2 inch guns.

An M 1897 French 75mm gun at Aberdeen Proving Ground in October 2009. Although not identical to the French 75mm guns in Irish service, this version of the French 75, modernized for use by U.S. forces, is quite similar to the Irish version supplied by the British. The barrel retains the "ears" at the muzzle end characteristic of the French 75. (Ralph Riccio)

One of the early Mk I 18-pdrs training in the Curragh, probably during The Emergency. The Mk I and Mk II guns that had equipped the Artillery Corps in 1923 were used throughout The Emergency, although by that time they had been updated with the Martin Parry carriage as shown here. (Lt Col Edmond Fogarty)

In 1943 the field artillery inventory consisted of eight 75mm guns, thirty-five (possibly thirty-seven) 18-pdrs, four 3.7 inch mountain howitzers, thirty-eight 4.5 inch howitzers, and six 60-pdrs for a total of 91-93 guns of all types. Based on a total of twenty-two field artillery batteries during The Emergency (twenty-one in the seven regiments, plus one, the 22nd, subordinate to Western Command), the requirement would have been 88 guns at four guns per battery, so at first glance, there would have been enough guns for each of the batteries, plus at least three guns to spare. However, the four 3.7 inch howitzers and the six 60-pdrs were assigned to independent batteries (1st Light, 41st and 42nd), resulting in only 81 guns available (75mm, 18-pdrs and 4.5 inch howitzers) for the twenty-two batteries, or a net shortfall of seven guns for the twenty-two batteries. Because the training establishment would have required some guns, it is problematic as to how many batteries actually had four guns each. By type, there would have been two 75mm batteries, eight 18-pdr batteries, and nine 4.5 inch batteries, for a total of nineteen batteries (76 guns), with five guns left over (three 18-pdrs and two 4.5 inch howitzers), bearing in mind that the 3.7 inch howitzers and 60-pdrs were assigned to independent batteries. Each of the seven field artillery regiments generally had at least one 18-pdr battery and one 4.5 inch howitzer battery, although the 2nd Regiment had two 75mm batteries and one 4.5 inch battery. Two of the regiments had two 4.5 inch batteries and one 18-pdr battery (4th and 5th Field Artillery Regiments); the remaining regiments were structured with two 18-pdr batteries and one 4.5 inch battery.

A restored 4.5 inch howitzer at McKee Barracks, Dublin, in October 2009. Britain supplied the Irish Army with some twenty-four 4.5 inch howitzers between 1941 and 1943, adding to those furnished earlier. (Sean Reilly, via Ian Stewart)

Equipment was assigned to the field artillery batteries as follows:

1st Field Artillery Battery: 18-pdr gun
2nd Field Artillery Battery: 18-pdr gun
3rd Field Artillery Battery: 4.5 inch howitzer
4th Field Artillery Battery: 4.5 inch howitzer
5th Field Artillery Battery: 18-pdr gun
6th Field Artillery Battery: 18-pdr gun
7th Field Artillery Battery: 18-pdr gun
8th Field Artillery Battery: 4.5 inch howitzer
9th Field Artillery Battery: 4.5 inch howitzer
10th Field Artillery Battery: 75mm gun
11th Field Artillery Battery: 4.5 inch howitzer
12th Field Artillery Battery: 18-pdr gun
13th Field Artillery Battery: 4.5 inch howitzer
14th Field Artillery Battery: 75mm gun
15th Field Artillery Battery: 4.5 inch howitzer
16th Field Artillery Battery: 18-pdr gun
17th Field Artillery Battery: 18-pdr gun
18th Field Artillery Battery: 18-pdr gun
19th Field Artillery Battery: 4.5 inch howitzer
20th Field Artillery Battery: 18-pdr gun
21st Field Artillery Battery: 4.5 inch howitzer
22nd Field Artillery Battery: 18-pdr gun
1st Light Battery: 3.7 inch mountain howitzer

A battery of 4.5" howitzers on a training exercise in Phoenix Park, Dublin, in May 1943. Crew members of the gun in the foreground are: No. 1, Sgt Brian Connolly, No. 2, Mickey Proctor, No. 3 John Doyle, No. 4 Larry Gleeson, No. 5, Patrick Killion. Mickey Proctor later made RSM and was extremely well regarded and respected in the Artillery Corps, having been described as the best soldier the Regiment ever had. (Shay Stone via Colin Stone)

This mortar crew are taking pains to carefully emplace this 120mm Brandt AM 50 mortar without its wheeled carriage. The Brandt, adopted in 1953, added considerable firepower and flexibility to Irish artillery units. (Mullingar Barracks Museum via Colin Stone)

Throughout the period from 1929 to 1945 training was taken very seriously and a high standard was achieved. In 1944 a divisional level artillery demonstration was held in the Glen of Imaal during which seventy-two guns (virtually all of the guns in the inventory at the time) were present and fired. This remains the artillery's largest shoot ever.

Post-War Organisation

In December 1946 the Proposed Peace Establishment disbanded four of the seven Field Artillery Regiments (the 3rd, 5th, 6th, and 7th), leaving the 1st, 2nd, and 4th Field Artillery Regiments. Most of the weapons of the disbanded regiments were mothballed into the Artillery Technical Stores in Kildare. Following the end of the war, the remaining artillery regiments were reduced from three batteries to two batteries, for a net loss of sixteen batteries. Paradoxically the number of guns in the regiments rose from twelve to sixteen, as the three four-gun batteries were replaced by two eight-gun batteries, so that reduction in the number of guns (from 84 in the seven wartime regiments down to 48 in the three post-war regiments) was not quite as drastic as were the organisational cuts. By 1949 the 25-pdr gun-howitzer had largely replaced the 18-pdr field gun that had been the mainstay of Irish artillery units since the Civil War in 1922. In the Eastern Command, the 2nd Brigade's 2nd Field Artillery Regiment in Dublin now consisted of the 10th and 14th Field Batteries, each equipped with eight 25-pdr Mk II gun-howitzers. Southern Command's 1st Brigade had the 1st Artillery Regiment in Ballincollig with the 1st and 2nd Field Batteries, also equipped with eight 25-pdrs each. Western Command counted the 4th Brigade's 4th Artillery Regiment in Mullingar, whose 8th and 15th Field Batteries also followed the standard eight-gun configuration, with 25-pdrs. By 1950, despite an adequate number of guns to equip all of the batteries, ammunition for the older guns (namely, the 18-pdr, 4.5 inch howitzer, and the 3.7 inch mountain howitzer) was no longer readily available. In 1953, artillery firepower was considerably bolstered by delivery of 72 Brandt M50A 120mm mortars.

An 18-pdr immediately after firing at the Glen of Imaal in 1960. The brass casing has barely cleared the breech block, and a gunner is kneeling with another round at the ready. The NCO on the far left is Sgt P. Mulcahy, later killed while serving in the Congo. (62 Reserve Artillery Regiment, via Capt Ian Stewart)

A battery of 25-pdrs drawn up for gun drill at Columb Barracks, Mullingar. Each gun is accompanied by an ammunition limber. The number of crew seems to vary from gun to gun. The tread pattern on the tires of the nearest gun are the WW II pattern used on guns in North Africa, attesting to fact that some of the 25-pdrs in Irish service had seen action in the desert. (Military Archives)

Beginning in about 1949, in addition to the 25-pdr guns previously mentioned, Ireland also acquired twelve 17-pdr and six 6-pdr anti-tank guns, which allowed anti-tank batteries to be organised for each of the three field artillery regiments then in being. Definitive information concerning the organisational structure and subordination of anti-tank batteries is sparse to nonexistent, but it seems probable that each of the three field artillery regiments had a six-gun anti-tank battery; the available guns would have allowed two batteries to be formed with 17-pdrs and one battery with 6-pdrs. The 4th Field Artillery Regiment had the 4th Anti-tank Battery, presumably from receipt of the anti-tank guns in 1949 until 1959, when it was disbanded; the battery likely was equipped with the 6-pdr guns. By extrapolation, the 1st and 2nd Field Artillery Regiments would have had similar anti-tank batteries, probably numbered 1st and 2nd, respectively, and equipped with the 17-pdr guns. Although there are indications that the 17-pdrs remained on hand until at least 1962, they too may have been removed from active service with the field artillery regiments in 1959, and the batteries they equipped disbanded. It is interesting that the 1959 organisation of the 1st Anti-aircraft Regiment (HQ Battery, one PDF AA Battery and three RDF AA batteries) had an anti-tank troop in each of the AA batteries for ground airfield defence. The AT troops in the RDF batteries were equipped with 17-pdr AT guns; the timing seems to indicate that the 17-pdrs were transferred from the field regiments to the anti-aircraft regiment.

In October 1959 the Army was reorganised by integrating the An Fórsa Cosanta Áitiúil (FCA) with the Permanent Defence Forces (PDF). The integration resulted in doubling both the number of artillery regiments (from three to six) as well as the number of batteries (from nine to eighteen). The artillery organisation was articulated on three PDF field artillery regiments (the 1st, 2nd, and 4th), located at Balllicollig, Dublin, and Mullingar, respectively, while three new field artillery regiments (the 3rd, 5th, and 6th) were formed to complement the three PDF field artillery regiments. The three FCA regiments were located at Templemore, Galway, and Dublin. Each regiment (PDF and FCA alike) consisted of three batteries; of the eighteen batteries, only three were PDF batteries (one in each of the PDF regiments); the remaining fifteen batteries were FCA batteries. Equipment varied from regiment to regiment, but included the venerable 18-pdrs, the newer 25-pdrs, the 4.5 inch howitzer, and the 120mm mortar. The battery numbers ran from 1st through 16th (excluding the 3rd), and 19th through 21st, but did not run consecutively within the regiments. Although the numbers of the new batteries resurrected those used during The Emergency, in fact the new batteries were based on infantry units whose personnel were retrained as artillerymen; the infantry units lost their names and took on traditional artillery numbers and designations.

By 1962 the 2-pdr and 6-pdr anti-tank guns had disappeared from the inventory, although the twelve 17-pdrs still appeared on hand. In the mid-1970s they were in deep care and preservation in Magee Barracks, Kildare, and when Magee Barracks closed in September 1998 the guns were distributed throughout the Army as gate guardians or general ornamentation. Once the 17-pdr anti-tank guns were boarded, the Artillery Corps was left with no anti-tank capability. However, since about 1959 various types of anti-armour weapons were issued to infantry units. The first of these was the Swedish 84mm Carl Gustav recoilless anti-tank weapon, introduced in the late 1950s, followed by the 90mm Bofors Pv 1110 introduced sometime in the 1960s. The Pv 1110 was withdrawn from service in the late 1990s, while as of 2012 the Carl Gustav remains in service with updated sights and ammunition; it is also in service with the Cavalry Corps, and there are two per battery with each artillery Field Battery as well. The French Milan anti-tank missile entered service initially with the artillery in the 1970s and later moved to the AT Platoon of the Support Company of the infantry battalions. In 1997 the Swedish AT-4 SRAAW (Short Range Anti-Armour Weapon) was acquired, and in 2003 the U.S. FGM-148 Javelin anti-armour weapons system made its debut in the Irish Army. The Javelin is the first anti–armour missile to be deployed with Irish peacekeepers in Chad (MINURCAT).

The 1979 Army reorganisation added three new FCA Field Artillery Regiments, these being the 7th, 8th, and 9th, bringing the total number of regiments to nine. There was only a modest asymetrical increase in the number of batteries, for a total of twenty-one batteries, which was quite close to the number that obtained during The Emergency. The only battery number missing from the 1st through the 22nd was the 3rd Battery. As in the previous organisation, only three batteries were PDF batteries. Each of the new regiments, as well as the three PDF regiments, were organised with a regimental headquarters, a headquarters battery, and two gun or mortar batteries; the 3rd, 5th, and 6th Regiments retained three batteries each. The 7th Field Artillery Regiment resurrected the number that designated one of the Field Artillery Regiments that had existed during The Emergency, and was based on the 14th and 19th Field Artillery Batteries (formerly of the 2nd Field Artillery Regiment); the 8th Field Artillery Regiment introduced a new number, and was based on the 2nd Field Artillery and 21st Heavy Mortar Batteries (ex 1st Field Artillery Regiment); and the 9th Field Artillery Regiment, also a new number, was based on the 15th and 20th Field Artillery batteries

120mm Brandt AM 50 mortars during firing exercises in the Glen of Imaal sometime in the 1960s. The classic mortar crew method of inserting fingers into the ears to protect against noise and concussion is evident and suggests that the mortar is about to be or has just been fired. (Brian Lowen)

A 25-pdr under tow by Bedford MFR2 6107 ZO at a parade in Newbridge sometime in the 1970s. (Denis McCarthy)

6-pdr anti-tank gun, Clancy Barracks Ordnance Stores, 11 September 1981. This is one of only six 6-pdrs acquired by Ireland in 1949. Anti-tank capability in the Irish Army during The Emergency was almost non-existent, limited to only seven 2-pdrs. (Denis McCarthy)

A Land Rover Defender 110 towing the 90mm Bofors Pv 1110 recoilless anti-tank weapon during a parade. (Sean O'Sullivan)

A 90mm Swedish Pv 1110 recoilless rifle as used by the Irish Army from the 1960s until the 1990s. This system was the most potent anti-tank system available to Irish infantry units at the time. (Denis McCarthy)

An FGM-148 Javelin at the moment of launch. The Javelin anti-armour weapon missile system is issued to infantry units and provides a potent anti-tank capability. (Irish Defence Forces)

Artillery corporal Brian Lowen at a plotting table on exercise in the field, late 1960s. Prior to the advent of computerized fire control for the guns, firing calculations were manually determined using aids such as plotting board shown. The plotting board is still used as backup in case the computers go down. (Brian Lowen)

(ex 4th Field Artillery Regiment), all equipped with 120mm Brandt mortars.

Based on the 2004 Reserve Defence Forces Implementation Plan, the Army's artillery organisation underwent a major restructuring. On 1 October 2005, the six FCA artillery regiments (3rd, 5th, 6th, 7th, 8th, and 9th) were consolidated into three new Reserve regiments, losing their FCA designation. The 3rd and 8th FCA Artillery Regiments merged to form the 31 Reserve Artillery Regiment; similarly, the 5th and 9th FCA Artillery Regiments merged to form the 54 Reserve Artillery Regiment, and the 6th and 7th FCA Artillery Regiments merged to form the 62 Reserve Artillery Regiment. The new regiments each consisted of three batteries, each of which was equipped with four 25-pdr guns

This photograph depicts two of the many interesting and colourful characters – from the 41st Battery – who have been associated with the Irish Artillery Corps. On the left is Gunner Sean Dunphy who represented Ireland at the 1967 Eurovision Song Contest and won second place with "If I Could Choose". He continued giving live performances until 2009, despite a quadruple bypass heart operation in 2007. He was a consummate gentleman, and at a funeral service for a Sergeant Major who had died after being retired 30 years, Dunphy, even though himself sick, sang a lament at the church, bringing the service to a hushed halt. He died in May 2011. On the right is Gunner Murphy, who is now retired. (Colin Stone)

A 25-pdr during the final 25-pdr shoot for the Irish Army on 30 July 2009. Judging from photographs, the crews who were selected to participate in this historic shoot were extremely enthusiastic and professional. Competition to be among the crews was fierce for this landmark event. (Colin Stone)

replacing the 120mm Brandt mortars that had previously equipped the FCA batteries, and which were phased out of service in 2004. In July 2009 the 25-pdrs in turn were phased out of the active inventory and were replaced by the newly acquired RUAG 120mm mortar.

62 RAR RUAGs firing in July 2009. The RUAG produces a pronounced muzzle blast, as evidenced by the mortar to the right which has just fired. (Colin Stone)

One of the two Nissan Pathfinders assigned to the Artillery School in the Curragh and fitted with the Field Artillery Computer Equipment (FACE) system. The Pathfinders replaced the earlier Landrover 109 in this role. The decal on the bonnet indicates that the vehicle belongs to the Combat Support College, which the Artillery School falls under organizationally. (Colin Stone)

Air-lifted artillery

On 28 June 1967, mortars and crews from the 19th Heavy Mortar Battery of the 2nd Field Artillery Regiment participated in a demonstration featuring the use of a helicopter in the artillery close support role. The 120mm AM 50 mortars and their crews were airlifted by two Air Corps SA 316B Alouette helicopters from their firing positions to a forward area where they resumed firing. This was the first occasion on which helicopters were used by the Defence Forces for this type of training. Subsequently, the Alouette was used occasionally to lift the 120mm mortar during exercises. The basic issue artillery piece of the period, the 25-pdr, was far too heavy to be airlifted by any of the helicopters in the Irish Air Corps inventory. In 1981 an SA 330 Puma was leased from Aerospatiale in order to test an aircraft to replace the SA 316 Alouette then in Irish service. During the testing the SA 330 was used to lift a 105 Light Gun with ease; ultimately, however, a decision was made to adopt the less capable SA 365F1 Dauphin, with the result that airlift of artillery remained restricted to lifting the 120mm Brandt mortar. It was not until the advent of the Agusta-Westland AW 139 in November 2006 that the Air Corps gained the capability to lift the 105mm Light Gun. Cooperative arrangements and procedures have been established between the Artillery Corps and the Air Corps, which has the necessary slings for airlifting artillery, and the AW 139 regularly lifts 105 Light Guns in exercises and demonstrations.

An Irish Air Corps Alouette III with 120mm Brandt mortar slung in a net. The photograph appeared on 29 June 1967, having been taken the previous day. This was the first instance in which Irish Air Corps helicopters were used to transport an Irish Artillery Corps mortar and crew. The mortar and crew were from the 19th Battery. (The Irish Times, via Brian Lowen)

An L118 Light Gun slung under an Irish Air Corps AW-139 helicopter during a demonstration at the Glen of Imaal in July 2009. The Artillery Corps and the Air Corps have developed joint procedures for airlifting the 105mm gun during exercises. (Irish Defence Forces)

An AW-139 helicopter with a105mm light gun during the occasion of the last 25-pdr shoot at the Glen of Imaal on 30 July 2009. Since the AW-139 entered service with the Irish Air Corps, it has afforded the Army an opportunity to rapidly deploy its gun and mortar assets.(Colin Stone)

Current Field Artillery Organisation

The field artillery organisation in early 2012 consisted of three PDF Field Artillery Regiments (1st, 2nd, and 4th) and three RDF Field Artillery Regiments (31, 54, and 62). The PDF regiments are equipped with the 105mm L118 Light Gun (one six-gun battery per regiment) and the 120mm RUAG mortar (one battery per regiment). The Artillery School has six 105mm L119 Light Guns and twelve 120mm RUAG mortars. The reserve regiments were equipped with the 25-pdr field gun (three four-gun batteries per regiment) until July 2009; upon retirement of the 25-pdrs at that time, the reserve batteries were re-equipped with the 120mm RUAG mortar. Thus, the PDF artillery regiments are equipped with a mix of field artillery and mortars, while the Reserve regiments are equipped exclusively with mortars. With the exception of six of the 105mm Light Guns that were delivered in 1980, all of the guns and mortars in the inventory, as of 2012, are fifteen years old or less. The 1st Regular and 31 Reserve Artillery Regiments, subordinate to Southern Brigade,

An L118 that has just been landed by an Agusta-Westland AW-139 multi-role helicopter during an exercise at the Glen of Imaal. The gun crews are also delivered by helicopter. (Lt Col Edmond Fogarty)

are located at McCann Barracks, Templemore; 2nd Regular and 62 Reserve Artillery Regiments, subordinate to Eastern Brigade, are located at McKee Barracks, Dublin; 4th Regular and 54 Reserve Artillery Regiments, subordinate to Western Brigade, are located at Columb Barracks, Mullingar; however, Columb Barracks are scheduled to be vacated on 30 March 2012.

Training with the 105mm L 118 light gun by the 6th Field Artillery Regiment, Glen of Imaal, 2004. Although the L118 is issued only to Regular Army field artillery batteries, reserve forces also train with them. (62 Reserve Artillery Regiment, via Capt Ian Stewart)

A 25-pdr during the final 25-pdr shoot on 30 July, 2009. Numerically, the 25-pdrs were the mainstay of the Irish artillery inventory from the late 1940s until they were phased out in 2009. (Irish Defence Forces)

Artillery in United Nations Operations

Artillery, in the form of 60mm, 81mm and 120mm mortars, has played its part in support of a number of Irish peacekeeping operations since 1960. An Artillery Mortar Troop of four 120mm AM 50 mortars has been deployed with every infantry battalion on peacekeeping operations since the first deployment overseas to the Congo, when the 1st Mortar Troop supporting 38 Infantry Battalion deployed there from November 1962 to May 1963. On 28 December 1962 the troop, firing in support of other U.N. forces, marked the first time that an Irish field artillery unit had been in combat. The Troop consisted of 42 all ranks drawn mainly from the 4th Field Artillery Regiment, and was equipped with 120mm Brandt mortars. The mortar troop was used in action in the Congo on several occasions. A mortar troop also deployed with the infantry groups to Cyprus. When the Defence Forces deployed to Lebanon with UNIFIL in 1978, this troop was supplemented with two Brandt 81mm mortars. The 60mm mortar was deployed with the Weapons Platoon of the infantry rifle companies, but never with the Artillery Mortar Troop/Platoon. The troop fired the 81s and 120s in action using illumination rounds. HE missions were commenced on several occasions, in particular when the unit was being shelled by the South Lebanese Army, which was an Israeli Defense Force backed militia.

An illumination round being readied for firing in an 81mm M44 mortar with the Irish UNIFIL contingent in Lebanon in 1980. The M44 was issued primarily to infantry units, but on occasion was also used by artillery units regularly equipped with the 120mm Brandt AM 50 mortar. (Martin Forde via Roy Kinsella)

The 120mm RUAG mortar was acquired from Switzerland in late 2008 and issued to Irish artillery units in mid-2009. The Irish Army's earlier 120mm mortar, the Brandt AM 50, had been phased out of service some four years earlier. (Irish Defence Forces)

A 120mm Brandt AM 50 mortar with the Irish contingent in Lebanon in 1980. Mortars were the only artillery available to the Irish contingent in Lebanon, and served well as a deterrent to attacks by the Lebanese militia. (Martin Forde, via Roy Kinsella)

A 120mm mortar supporting the Irish Battalion, UNIFIL, returning fire at night against Lebanese militia. Illumination rounds were commonly used to reveal enemy positions. (Lt Col Edmond Fogarty)

The Orbiter UAV – with prominent Irish markings – as deployed in Chad. The Orbiter has a 2.2 meter wingspan, weighs 14.3 pounds, and can be operated remotely by an operator as far as 15 km distant. (RSM Mick Dillon)

The fire missions usually were terminated once the Command Post reported ready to the OP, as the ready message had been intercepted by the South Lebanese Army/Israeli Defense Force who thus knew that the Irish mortar troop were about to commence firing on one of their positions. The final rounds fired by the Irish UNIFIL contingent occurred in 2001 in the Bekaa Valley. The Mortar Troop deployed to Chad in 2009 was equipped with the 81mm LMLB mortar, and more recently, in 2011 the RUAG 120mm mortar was deployed to support 105 Infantry Battalion in UNIFIL.

An interesting and noteworthy addition to the Irish Artillery Corps equipment inventory is the Israeli Orbiter unmanned aerial vehicle (UAV). The army adopted the Orbiter for use as a reconnaissance platform in Chad, ordering two systems in 2007. Each system consists of three aircraft fitted with day/night 10x zoom color video camera and a controller, plus batteries and a lighter daytime only camera. The Orbiter is built by ADS (Aeronautics Defence systems, Ltd.); each system cost €390,000, or $550,000.00 USD each, for a total cost of €780,000.

Although Irish troops like the system and its capabilities, it has suffered a high attrition rate. At least one UAV was lost in Chad on 29 December 2008 when the Orbiter lost its communications link with the operator and then defaulted to a GPS location stored in its memory and tried to fly back to Ireland, some 5,000 km away – predictably, it ran out of power after about 200 km and was lost somewhere in the desert.

The Orbiter being prepared for flight in Chad; it is launched by catapult and lands using a parachute and an air bag. The UAV is battery powered and can stay aloft up to three hours when fitted with the lighter daylight only camera. The Nissan Patrol in the background with the cage on top is used to transport the UAV. (RSM Mick Dillon)

An operator with the control unit display in Chad. The Orbiter can fly as high as 3,000 meters but usually is employed at altitudes between 200 to 1,500 meters. Maximum speed is 135 km/h, but it normally is flown between 45-120 km/h. (RSM Mick Dillon)

Artillery Unit Name and Numbering Conventions

Although Irish field artillery and air defence numbering conventions generally are simple, straightforward and consistent, there are some variations and nuances that bear explanation.

As a general rule, field and air defence artillery units are or have been sequentially numbered, beginning with the number 1. Historically, field artillery brigades bore the numbers 1, 2, and 4; battalions 1 through 4; and batteries 1 through 22 (5 through 22 being originally assigned to Volunteer Force batteries). Battery numbers 23 through 28 were assigned to Volunteer Force units, but were not subsequently used for active units. Some duplicate numbers were encountered for the Light Batteries: 1st Light Battery was a Regular unit, and 2nd through 7th Light Batteries were Volunteer Force units, which in actuality were mortar batteries. Likewise, air defence units duplicate some of the numbers of the field artillery units: 1st Air Defence Regiment has subordinate to it the 1st through 4th Air Defence Batteries.

For a period between 1944 and 1959 several independent artillery batteries in the 40 series existed (41st, 42nd, 44th, 45th, and 46th), but these eventually were redesignated with numbers from the 1 through 22 range.

Except for the three Reserve Artillery Regiments, the unit numbers are expressed in ordinal rather than cardinal numbers (i.e., 1st Field Artillery Regiment, etc.). Artillery regiments and batteries are always referred to by Arabic numerals, whereas the brigades and battalions can at times be seen referred to by Roman numerals and expressed as cardinal rather than ordinal numbers (i.e., IV Field Artillery Brigade, or II Field Artillery Battalion).

From the inception of the Artillery Corps in 1923 until 1943 artillery batteries were designated simply as field batteries, and in 1943 the designation was changed to field artillery batteries. In 1959 artillery batteries that were equipped with mortars had their designations changed to heavy mortar batteries to reflect the type of equipment (120mm mortars) issued. Also, during the 1959 integration of FCA and PDF units, the new FCA artillery batteries were not formed *ex novo*, but instead were created by re-roling infantry units as artillery units, dropping the former infantry unit names and adopting the numbers of artillery batteries that had been disbanded during the 1945-46 Army reorganisation.

In 1998 the batteries in the three PDF regiments (1st, 2nd, and 4th) were renumbered as 1, 2, and 3 Batteries in each of the regiments, replacing the traditional battery numbers that had been retained since 1923. In 2005, when the FCA regiments were reorganised and formed into 31, 54, and 62 Reserve Artillery Regiments, the old battery numbers likewise were replaced by the Numbers 1, 2, and 3 Gun Battery in each of the new Reserve regiments. The new numbering system, while undeniably introducing an element of uniformity and standardization throughout the regiments, has led to a complete loss of easily recognizable historical identity and continuity.

The new battery numbers are now expressed as cardinal rather than ordinal numbers, as are the numbers for the reserve regiments. Another note with respect to the reserve regiments is that their two-digit designators properly are read as "Three One," "Five Four," and "Six Two." The first digit of each number refers to one of the old FCA regiments that formed the new regiment, and the second number indicates both the number of the brigade to which it is subordinate, as well as, coincidentally, the number of the Regular artillery regiment with which it is associated or twinned.

The coast artillery batteries basically were identified by the name of the fort that they were located in (i.e., Fort Davis, Fort Mitchel, Dunree, etc.). Bere Island, which had a number of separate and distinct battery locations, identified the batteries by specific location name (i.e., Lonehort Battery).

Artillery Traditions

All military organizations have unique traditions, and the Irish Artillery Corps is no exception to the rule. The Glen of Imaal training area, although surely often cursed in eloquently creative phrases that only the Irish can coax out of the English language, has a special place in the hearts of the gunners. One tradition was for artillerymen to sign their names on the walls of the old WW II era wooden billets there; in some instances, sons were able to find space to sign where their fathers had signed long before them. Unfortunately, the walls have now been clad over, but the signatures remain. Another custom is to honour dead gunners by firing a shell with the deceased's name and regiment on it as a mark of respect. One gunner, currently serving as this book is being written, annually fires a round inscribed with his family's name as a tribute to deceased and past serving members of the extended family, many of whom served in the Artillery Corps. As a variation of the practice, the ashes former Bty. Sgt. George Keogh, according to his wishes, were shipped from Australia and scattered over the Glen by the 14th Battery during annual training in 1989. After the ashes were spread, songs were sung in Leitrim gun position prior to adjourning to Fenton's Pub where Keogh was given a proper artilleryman's farewell.

Fenton's Pub, in Donard, Co. Wicklow, is a mecca for artillerymen participating in shoots in the Glen of Imaal. This photo from 1990 shows Don Fenton, the owner (with mustache), and Gunner Gerry Corcoran (with his hand on his ear), then a driver with the 6th Field Artillery Regiment PDF and as of late 2009 still with the 62 Reserve Artillery Regiment. The walls are hung with numerous artillery-related pictures. The pub originally was known as Bob Allen's, and the Cork artillery regiments still refer to Fenton's as Bob's. (Captain Ian Stewart)

Field Artillery Brigades, Battalions, Regiments and Batteries

1st Field Artillery Brigade: The 1st Field Artillery Brigade was formed in 1931 and consisted of the 1st, 2nd, and 3rd Field Batteries, with 18-pdrs, and the 4th Field Battery with the 4.5 inch howitzer. In essence, the brigade was the parent unit for all four Regular artillery batteries until about 1940, when the artillery battalions were formed. The relationship of the 1st Brigade to the 1st Battalion is not entirely clear, but it appears that the Brigade was transformed into or renamed as the 1st Field Artillery Battalion in 1940, which maintained a somewhat similar organisational structure as the Brigade.

2nd Field Artillery Brigade (Reserve): The 2nd Field Artillery Brigade was formed in April 1939 at McKee Barracks. The brigade comprised four field batteries (5th, 6th, 7th, and 8th) and a mortar battery (the 2nd). The 5th Field Battery was a Regular battery, while the remaining batteries were all Volunteer Force batteries. Upon outbreak of war, the brigade moved to its wartime location in Mullingar. Three batteries were accommodated under canvas at Mullingar Barrracks at what was then the green portion of Barracks Square (later replaced by the dining facility), the 5th Field Battery and 2nd Mortar Battery took over Knockdrin Castle. The Volunteer Forces gunners were amalgamated into the brigade, and in June 1940 the mortars were transferred to the infantry. In June 1940 the brigade was reorganised with two subordinate battalions, the 2nd and 4th, and stationed in Collins Barracks, Dublin

4th Field Artillery Brigade (Reserve): Like the 2nd Field Artillery Brigade, the 4th Field Artillery Brigade controlled a number of Volunteer Force batteries. Although based only on anecdotal evidence, it appears that the 13th, 14th, and 16th Field Batteries, plus the 4th Light Battery, were subordinate to the 4th Brigade.

1st Field Artillery Battalion (Cork): The 1st Field Artillery Battalion consisted of the 1st, 2nd, 3rd, and 4th Field Batteries, as well as of the 1st Light Battery. At some point, the 7th Field Battery was also subordinate to the 1st Battalion. The 1st Field Artillery Battalion appears to have been a successor to the 1st Field Artillery Brigade, probably inheriting the Brigade's subordinate batteries in early to mid-1940. Upon formation of the field artillery regiments in 1943 the battalion's batteries were resubordinated to various regiments.

2nd Field Artillery Battalion (Dublin): The 2nd Field Artillery Battalion, formed in June 1940 was one of the two battalions of the 2nd Field Artillery Brigade, and consisted of the 6th, 10th, and 11th Field Batteries. About a year later, around June 1941, the 4th Field Artillery Battalion was merged with the 2nd F.A. Battalion, and the amalgamated battalion, designated the 2nd Field Artillery Battalion, consisted of the 5th, 6th, 7th, 8th, 10th, and 11th Field Batteries. In 1943 the 2nd Field Artillery Battalion formed the basis of the 2nd Field Artillery Regiment, as well as of the 4th and 6th Field Artillery Regiments.

A French 75 assigned to the 41st/42nd Battery in August 1950. The wheels of the gun are resting on a circular firing platform of the type used by the 25-pdr. The members of the crew are: Sgt J. Kirwen, Cpl J. Hazley, Cpl C. Smith, Gnr M. O'Beirne, Gnr F. Scalley, Gnr T. O'Sullivan, and Gnr T. McMahon. (Shay Stone via Colin Stone)

3rd Field Artillery Battalion: The 3rd Field Artillery Battalion consisted of the 4th, 9th, and 12th Field Batteries, and at least for training purposes, also controlled the 1st and 2nd Field Batteries sometime during 1941. Date of formation of the 3rd Battalion is uncertain, but likely was formed in June 1940 along with the 2nd and 4th Field Artillery Battalions. Upon formation of the field artillery regiments in 1943, the battalion's batteries were resubordinated to various regiments.

4th Field Artillery Battalion (Mullingar): Along with the 2nd Field Artillery battalion, the 4th Field Artillery Battalion was formed in June 1940 and consisted of 5th, 7th, and 8th Field Batteries. In or around June 1941 its battalions were amalgamated with those of the 2nd Field Artillery Battalion, resulting in an expanded 2nd Field Artillery Battalion.

1st Field Artillery Regiment (McCann Barracks, Templemore): The 1st Field Artillery Regiment was the first artillery regiment formed in the Irish Army, in Clonmel in late 1943, based on the 1st, 2nd, and 21st Field Batteries. In October 1945 the regiment, along with the 7th Field Artillery Regiment, moved from Clonmel to Murphy Barracks, Ballincollig for the purpose of demobilisation. At the conclusion of the demobilisation process, which occurred over a period of twelve months and during which the 21st Battery was eliminated, the remaining members of the three regiments combined to form the new 1st Field Artillery Regiment. The regiment was equipped with 18-pdrs, but in 1949 the regiment was re-equipped with the 25-pdr that had just been acquired by Ireland, and battery organisation had changed from a four-gun to an eight-gun battery. In 1959, the regiment was reorganised to comprise three gun batteries: the 1st Battery was a regular battery, equipped with the 25-pdr, while the 2nd and 21st Batteries were FCA batteries, equipped with the 4.5 inch howitzer and the 120mm mortar, respectively. The regiment is now subordinate to 1st Southern Brigade, and is located at McCann Barracks, Templemore, and has one six-gun battery of L 118 Light Guns, as well as one mortar battery equipped with the 120mm RUAG mortar. The regiment also maintains six guns as equipment only for a 25-pdr ceremonial battery.

2nd Field Artillery Regiment (McKee Barracks, Dublin): Formally constituted as the 2nd Field Artillery Regiment in 1943, the parentage of the regiment can be traced back to 1934 and the Volunteer Force, which included the 1st Field Artillery Brigade at Plunkett Barracks, with three batteries. In 1939, with the war imminent, plans were made to organise a new artillery brigade, the 2nd, with four field batteries and a mortar battery. In 1940 the brigade was downgraded to battalion status; the 2nd Field Artillery Battalion was stationed at Collins Barracks, Dublin. In 1943 the 2nd Field Artillery Battalion formed the basis of the 2nd Field Artillery Regiment, as well as of the 4th and 6th Field Artillery Regiments. The regiment consisted of three batteries (10th, 14th, and 19th), plus an anti-tank battery with 2-pdr guns (the anti-tank battery likely had no more than two guns). The 10th and 14th Batteries were the only batteries in the Army equipped with the 75mm field gun; the 19th Battery had four 4.5 inch howitzers. Following the drastic cutbacks to the Army in 1946, the regiment was parceled out between three barracks in Dublin, namely, Cathal Brugha, Griffith, and Collins Barracks. In 1959, with the integration of the PDF and

A 4.5 inch howitzer firing at the Glen of Imaal, probably in the 1960s. The 4.5 inch howitzer was the heaviest standard field piece in the Irish Army. Generally speaking, each field artillery regiment had one battery of 4.5 howitzers. (Sean Murphy)

Two newly acquired ACMATS belonging to 2nd FAR, towing nicely tarped 25-pdrs. The photograph dates to 1985 and was taken while the guns were en route to the Glen of Imaal, via Kilcullen, as shown on the road sign. (Sean O'Sullivan)

Two 105mm light guns and Bofors at an exhibition of military equipment on 4 May 1983. The gun in the foreground has its barrel in the travel position, while the gun in the middle has its barrel at maximum elevation. The elevation of the second gun's barrel and that of the 40mm Bofors L60 anti-aircraft gun next to it makes an interesting comparison. At the time of this photograph, these two L118 guns represented one third of that type of gun in the Irish inventory. (Denis McCarthy)

The 2nd Field Artillery Regiment's 25-pdr ceremonial battery firing a 21 gun salute during the inauguration of Mr. Michael D. Higgins as President of Ireland, 11 November 2011. (Irish Defence Forces)

FCA, the 2nd and 7th Field Artillery Regiments were amalgamated and the artillery assets in the Dublin area were concentrated at Collins Barracks and McKee Barracks. In 1982 the 2nd Field Artillery Regiment became the first Irish artillery regiment to be equipped with the new 105mm L118 Light Gun, replacing the 25-pdrs. The regiment is now subordinate to 2 Eastern Brigade, and has one battery of six L 118 Light Guns, and two 120mm mortar batteries. The regiment also holds six guns for a 25-pdr ceremonial battery, but the guns are maintained as equipment only and are manned by crews only for ceremonial occasions.

3rd Field Artillery Regiment (McCann Barracks, Templemore): The 3rd Field Artillery Regiment was one of the seven artillery regiments active during The Emergency period, and in December 1946 it was one of the four artillery regiments to be disbanded. On 1 October, 1959, upon integration of the FCA with the PDF, the 3rd Field Artillery Regiment (FCA) was formed. The resurrected regiment consisted of the 9th Field Artillery Battery (Templemore) with 4.5 inch howitzers; the 12th Field Artillery Battery (Birr) equipped with the 4.5 howitzer; and the 16th Heavy Mortar Battery (Thurles) with the 120mm mortar. In 1978 the equipment of the 9th Battery was upgraded from the 4.5 inch howitzer to the 25-pdr field gun, and the 12th Battery gave up its 4.5 inch howitzers, drawing four 120mm mortars from the 16th Battery; the 16th Heavy Mortar Battery was reduced from eight to four 120mm mortars. In October 2005 the regiment was merged with the 8th Field Artillery Regiment to become part of 31 Reserve Artillery Regiment.

4th Field Artillery Regiment (Columb Barracks, Mullingar): The 4th Field Artillery Regiment was formally stood up on 17 February, 1943, based on the former 2nd Artillery Battalion. In December 1946 it was amalgamated with the 5th Field Artillery Regiment, with the 20th Battery being disbanded. The resulting regiment consisted of the 8th and 15th Field Artillery Batteries and the 4th Anti-Tank Battery. In 1959 the anti-tank battery was disbanded, the 15th Battery became an FCA battery, and the 20th Battery was reactivated as an FCA battery. On 18 May, 1998, the regiment received a battery of the 105mm L118 Light Gun. As of 2012 it constitutes the artillery regiment of 4th Western Brigade, Athlone. It is equipped with one six-gun battery of L118 105mm Light Guns, and one 120mm mortar battery. The regiment is scheduled to move from Columb Barrracks upon its closure scheduled for 30 March 2012 to Custume Barracks, Athlone.

An ACMAT VLRA TPK 420 with hard top cab, taken at the Curragh Military Show in May 2009. This vehicle is now owned privately; it was part of the fourth and final batch of VLRAs bought by the Irish Army in 1993, and with the addition of a SINCGARS radio was converted to a 4th FAR command post around 2000. Note the radio antenna next to the passenger door. The hard top version of the ACMAT was issued only to 4th FAR. (Colin Stone)

5th Field Artillery Regiment (Galway): The 5th Field Artillery Regiment was one of the seven artillery regiments active during The Emergency period. In December 1946 it was one of the four artillery regiments to be disbanded, and was amalgamated with the 4th Field Artillery Regiment. Upon integration of the FCA with the PDF in October 1959, the 5th Field Artillery Regiment (FCA) was formed 30 October 1959 with a nucleus of officer and NCO staff from the 4th Field Artillery Regiment, and consisted of the 4th Battery (ex-50th Battalion, Galway) with 25-pdrs; the 7th Battery (ex-Gort Battalion) with120mm mortars, and the 13th Heavy Mortar Battery (ex-Tuam Battalion), also with 120mm mortars. Merged with the 9th Field Artillery Regiment, it became part of 54 Reserve Field Artillery Regiment in 2005.

6th Field Artillery Regiment (Kildare): The 6th Field Artillery Regiment was one of the seven artillery regiments active during The Emergency period. In December 1946 it was one of the four artillery regiments to be disbanded. The regiment was reconstituted as an FCA regiment in the 1959 reorganisation, with the 5th, 6th, and 11th Batteries, and with headquarters at Magee Barracks, Kildare. In October 2005 the regiment was merged with the 7th Field Artillery Regiment to become part of the 62 Reserve Artillery Regiment, and its batteries formed the basis of the No. 2 and No. 3 Gun Batteries of 62 Regiment.

7th Field Artillery Regiment (Limerick): The 7th Field Artillery Regiment was one of the seven artillery regiments active during The Emergency period, and in December 1946 it was one of the four artillery regiments to be disbanded. In 1979 a new 7th Field Artillery Regiment was formed in Dublin, incorporating the 14th and 19th Batteries, and in October 2005 it was merged with the 6th Field Artillery Regiment to become part of the 62 Reserve Artillery Regiment, and its batteries formed the basis of the HQ Battery and No. 1 Gun Battery of 62 Regiment.

8th Field Artillery Regiment (Murphy Barracks, Ballincollig): The precursor units of the 8th Field Artillery Regiment were the independent 44th and 45th Field Artillery Batteries, which became the 2nd Field Artillery Battery

AA 120mm Brandt mortar firing at Stranahealy. The mortar detachment are wearing the recently introduced DPM uniform and are also wearing ear protection devices – a significant improvement over the older method of simply sticking index fingers into the ears to attenuate the sound of firing. (62 Reserve Artillery Regiment, via Capt Ian Stewart)

An ACMAT towing a 25-pdr gun of the 6th Field Artillery Regiment, Glen of Imaal, in 1989. Both the breech and muzzle of the gun have been covered while in transit. (62 Reserve Artillery Regiment, via Capt Ian Stewart)

The last 25-pdr shoot for 14th Battery, 7th FAR, 12 July 2005. The No. 2 is one of a number of females serving in the Artillery Corps. In 1990 the Irish Army began to allow females to serve in selected appointments in the Reserves, and in 1992 the practice spread to all arms. (Colin Stone)

Elements of 1st and 8th FAR leaving Murphy Barracks, Ballincollig, in 1998 upon closure of the barracks. As with any such move, the occasion was somber rather than festive. (Sgt Mick Royanne via Colin Stone)

and the 21st Heavy Mortar Battery, respectively, of the 1st Field Artillery Regiment in 1959, in turn becoming part of the 8th Field Artillery Regiment in 1979. In October 2005 the 8th Regiment was merged with the 3rd Field Artillery Regiment to become part of the 31 Reserve Artillery Regiment.

9th Field Artillery Regiment (Columb Barracks, Mullingar): The 9th Field Artillery Regiment was formed as a new regiment in 1979, based on the 15th and 20th Field Artillery Batteries. In October 2005 the regiment merged with the 5th Field Artillery Regiment and formed the 54 Reserve Field Artillery Regiment.

31 Reserve Artillery Regiment (McCann Barracks, Templemore, Co. Tipperary): 31 Reserve Artillery Regiment was formed on 1 October 2005, based on reorganisation of the old FCA. Two former FCA units, the 3rd FCA Field Artillery Regiment (Tipperary) and the 8th FCA Field Artillery Regiment (Ballincollig), as well as A Coy 14th Infantry Battalion (Nenagh) were disbanded, their equipment and personnel then forming the new 31 Reserve Regiment. The regiment has three batteries: No. 1 Gun Battery (Collins Barracks, Cork); No. 2 Gun Battery (Thurles, Co. Tipperary); No. 3 Gun Battery (Nenagh, Co. Tipperary). The regiment is subordinate to 1 Southern Brigade RDF. The gun batteries were equipped with four 25-pdr field guns until July 2009, when the 25-pdrs were removed from service, to be replaced by the newly acquired RUAG 120mm mortar.

A mortar detachment setting up a 120mm RUAG mortar for firing. The RUAG is towed on a trailer, but then set up on a conventional base plate and bipod for firing; the earlier 120mm Brandt AM 50 in Irish service had an integral trailer that remained attached to the mortar for firing. (Irish Defence Forces)

Laying a 120mm RUAG mortar. The RUAG, which began entering service in Ireland in 2008, in terms of numbers of tubes, outnumbers the 105mm L118 and is now essentially the mainstay of the equipment inventory in the Irish Artillery Corps. (Colin Stone)

A partially camouflaged 25-pdr taken at camp in 2008. This particular gun, No. 19, was on loan to 62 RAR from 54 RAR's Galway battery for the annual shoot. The Galway guns had their unit logo painted on the shield rather than on a decal. (Colin Stone)

54 Reserve Artillery Regiment (Columb Barracks, Mullingar, Co. Westmeath): 54 Reserve Artillery Regiment was formed on 1 October 2005 from the 5th and 9th FCA Field Artillery Regiments. The regiment has three batteries: No. 1 Gun Battery (Gort, Co. Galway); No. 2 Gun Battery (Tuam, Co. Galway); No. 3 Gun Battery (Columb Barracks, Mullingar, Co. Westmeath). It is subordinate to 4th Western Reserve Brigade. The gun batteries were equipped with four 25-pdr field guns until July 2009, when the 25-pdrs were removed from service, to be replaced by the newly acquired RUAG 120mm mortar. The relocation of HQ Battery and No. 3 Gun Battery upon closure of Columb Barracks scheduled for 30 March 2012 is unclear.

62 Reserve Artillery Regiment: (McKee Barracks, Phoenix Park, Dublin): 62 Reserve Artillery Regiment was formed on 1 October, 2005 by the amalgamation of the 6th and 7th FCA Field Artillery Regiments. The regiment has three batteries: No. 1 Gun Battery (McKee Barracks, Dublin); No. 2 Gun Battery (Connolly Barracks, Curragh Camp, Co. Kildare); No. 3 Gun Battery (Connolly Barracks, Curragh Camp, Co. Kildare). It is subordinate to 2nd Eastern Reserve Brigade. The gun batteries were equipped with four 25-pdr field guns until July 2009 when the 25-pdrs were removed from service, to be replaced by the newly acquired RUAG 120mm mortar.

1st Field Artillery Battery (Collins Barracks, Cork): The 1st Field Artillery Battery, as its name implies, was the first field artillery battery to be formed in the Artillery Corps. It was formed in Dublin on 10 July 1923, designated as the 1st Field Battery, and was equipped with four Mk II 18-pdrs. In 1928 the battery was moved to Kildare, but was assigned to support the 4th Infantry Battalion stationed in Cork. This was not an ideal arrangement, as it meant that normally the battery was obliged to load onto trains for a day's journey to Cork; on one occasion, in 1936, it took eight days to move by road from Kildare to Cork, as the guns were still horse-drawn. In 1931, the battery was one of three batteries of the 1st Field Artillery Brigade. From April 1934 until April 1939, the battery was responsible for training volunteers from the Southern Command. In 1939, the 1st Battery was issued newly modernized 18-pdrs fitted with rubber tired carriages and, despite initial crew misgivings prior to testing, set the standard for a fairly quick changeover from horse-drawn to mechanically towed guns. By April 1940, the battery had become part of the 1st Field Artillery Battalion, and in 1941 was assigned to the 10th Infantry Battalion in Templemore. In 1943 artillery assets were grouped into regiments, and the 1st Field Artillery Battery became subordinate to the 1st Artillery Regiment. By 1949, the battery had relinquished its 18-pdrs and was equipped with six of the recently acquired 25-pdr field guns. In 1998 the battery was reequipped with six 105mm L118 Light Guns. The 1st Field Artillery Battery has throughout the years maintained its status as one of the premier batteries in the Irish Artillery Corps establishment, being retained as a Regular or Permanent unit and receiving priority for issue of the latest field artillery equipment. Since 1940, it has remained subordinate to the 1st Artillery Regiment. In 1998 this battery was redesignated as No. 1 Gun Battery, 1st Field Artillery Regiment, and moved from Murphy Barracks, Ballincollig, to Collins Barracks, Cork. It is the only battery in the Irish Artillery Corps to have retained its original number continuously from 1923 through the various reorganisations, including the 1998 renumbering of the PDF artillery batteries.

2nd Field Artillery Battery (Dublin): The 2nd Field Artillery Battery was formed in Dublin on 5 January 1925 at McKee Barracks, Dublin, was designated as the 2nd Field Battery, and was equipped with four Mk I 18-pdrs. In 1928, along with the 1st Battery, it was moved to Kildare and assigned to support the Dublin based 5th Infantry Battalion. In 1931, the battery was one of three batteries of the 1st Field Artillery Brigade, but was in a reserve status. From 1934 to 1939, the 2nd Field Battery was responsible for training volunteers from the Eastern Command. In 1936 the battery's gun equipment was updated with the newer box-trail Mk IV 18-pdr replacing the old Mk I guns. In 1937 the 2nd Battery was relegated to the status of a reserve battery, but in 1939, with the outbreak of war in Europe, the battery was once again brought to active status and assigned to Fort Duncannon in County Wexford to defend Waterford Harbour. In 1941 the battery was assigned to support the 13th Infantry Battalion in Cahir; and in 1941 the battery also trained on the 4.5 inch howitzer. In 1943 the battery was moved to Clonmel, along with the 1st Battery and the newly formed 21st Battery, forming part of the 1st Field Artillery Regiment. During The Emergency period the battery was equipped with four 18- pdrs; at some point it appears that the 18-pdrs may have been replaced by the 4.5 inch howitzer, as according to one source the last 4.5 inch round was fired by the 2nd Battery on 26 July 1977. In 1977 equipment was upgraded to 25-pdrs. The battery was subordinate to the 1st Field Artillery Regiment from 1943 through until 1959, when the independent 44th Field Artillery Battery (FCA) was renamed as the 2nd Field Artillery Battery. In 1979, the battery became part of the newly formed 8th Field Artillery Regiment. On 1 October 2005 it merged with the 21st Heavy Mortar Battery to become No. 1 Gun Battery of 31 Reserve Artillery Regiment.

105mm light gun at Cathal Brugha Barracks, 31 August 1998, at a display marking the 50th anniversary of the "End of The Emergency". Between 1980 and 1998 the Irish Army had only one battery of the L118, and in 1998 received enough guns to form a further three batteries. (Denis McCarthy)

Breech of 18-pdr at Curragh Camp. The markings indicate it as a QF 18 Pr Mk IV, serial number 15714, manufactured in 1920. The Irish Army's 18-pdr inventory consisted of a mix of the early Mk I and Mk II models, as well as the later Mk IV model. (Ralph Riccio)

A 2nd FAR battery of 25-pdrs drawn up on parade; the ACMAT gun towers obscure most of the guns themselves. Land Rovers are in the foreground, and a Bedford MFR 2 is behind the ACMATS and guns. (RSM Mick Dillon via Colin Stone)

3rd Field Artillery Battery: The 3rd Field Artillery Battery was formed in 1925 and was equipped with four Ordnance QF 4.5 inch howitzers. In 1931, the battery was one of three batteries of the 1st Field Artillery Brigade. Between 1934 and 1939, the 3rd Field Battery was responsible for training volunteers from the Western Command. In 1936 it was stationed in Athlone, and during The Emergency it was subordinate to the 7th Field Artillery Regiment. Following demobilization in 1946, the battery was disbanded, and reappeared in 1979 when it was subordinate to the Depot and School, Artillery Corps with six 105mm L118 Light Guns, on the equipment scale of the Artillery School.

4th Field Artillery Battery (Galway): The 4th Field Battery probably was formed in 1933 and was part of the 1st Field Artillery Brigade. It was equipped with four 4.5 inch howitzers, and was an "A" Reserve unit. During The Emergency it was subordinate to the 5th Field Artillery Regiment, but was demobilized along with the 5th Regiment in 1946. In 1959, with the integration of the PDF and the FCA, the 50th Infantry Battalion, Galway was re-roled as an artillery battery, becoming the 4th Field Artillery Battery, subordinate to the new 5th Field Artillery Regiment (FCA) and was equipped with six 25-pdr guns. The battery lost its historical identity and effectively ceased to exist when the 5th and 9th Regiments merged in October 2005 to form the 54 Reserve Artillery Regiment, and the resulting batteries of the new regiment were assigned new numbers (1, 2, and 3).

5th Field Artillery Battery (Kildare): Originally a Volunteer Force field battery based in Dublin in 1936, by 1940 the battery was subordinate to 4th Artillery Battalion in Western Command. In 1941 it was subordinate to the 2nd Field Artillery Battalion and was based in Mullingar. During The Emergency this battery was subordinate to the 6th Field Artillery Regiment and was equipped with four 18-pdr guns. As part of the 6th Field Artillery Regiment, it was disbanded during the 1945 demobilisation. In 1959 the South Kildare Infantry Battalion became the 5th Field Artillery Battery of the new 6th Field Artillery Regiment (FCA) and was equipped with eight 18-pdrs. When the 6th and 7th Regiments merged in October 2005 to form the 62 Reserve Artillery Regiment the 5th Battery, along with the 6th and 11th Batteries of the 6th Regiment, formed the No. 2 and No. 3 Gun Batteries of the new 62 Regiment.

6th Field Artillery Battery (Naas): Originally a Volunteer Force field battery based in Dublin, by 1940 the battery was subordinate to 4th Artillery Battalion in Western Command, and was attached to the 2nd Infantry Battalion, Ballsbridge. In 1941 the battery trained on the 4.5 inch

A 6th Field Artillery Regiment 25-pdr crew taking a break during an exercise at the Glen of Imaal. The absence of ammunition near the gun and the fresh tire tracks made by the gun tower likely indicate that the gun has just been deployed prior to firing. (62 Reserve Artillery Regiment, via Capt Ian Stewart)

howitzer. During The Emergency the battery, subordinate to the 6th Field Artillery Regiment, was equipped with four 18-pdr guns, and was demobilized in 1945. In October 1959 the North Kildare Infantry Battalion became the 6th Field Artillery Battery (FCA), with headquarters at Naas and subordinate to the newly formed 6th Field Artillery Regiment (FCA), and was equipped with eight 18-pdrs. In 1977 it was re-equipped with six 25-pdr guns. Throughout its operational life it remained subordinate to the 6th Field Artillery Regiment. When the 6th and 7th Regiments merged in October 2005 to form the 62 Reserve Artillery Regiment, the 6th Battery, along with its 5th and 11th sister batteries of the 6th Regiment, formed the No. 2 and No. 3 Gun Batteries of the new 62 Regiment.

A pair of 25-pdrs on exercise in the Glen of Imaal in 1978. These guns were assigned to 6th Field Artillery Regiment. The yellow projectiles in the foreground are high explosive rounds being readied for firing. (62 Reserve Artillery Regiment, via Capt Ian Stewart)

7th Field Artillery Battery (Gort): Originally a Volunteer Force field battery based in Monaghan, and in 1941was stationed at McKee Barracks, Dublin, as part of the 2nd Field Artillery Battalion. The 7th Field Artillery Battery was subordinate to the 5th Field Artillery Regiment during The Emergency, and was demobilised when the Army was reduced in size in 1945-46. During The Emergency, it was equipped with four 18-pdr guns. In 1959 the 7th Field Artillery Battery was reborn, based on the Gort Infantry Battalion, FCA, and subordinate to the newly formed 5th Field Artillery Regiment (FCA). The metamorphosis from infantry to artillery was completed by issue of the 120mm Brandt mortar to the battery. When the 5th and 9th Regiments merged in October 2005 to form the 54 Reserve Artillery Regiment, the 7th Battery became the basis for No. 1 Gun Battery of 54 Reserve Artillery Regiment.

8th Field Artillery Battery: Originally a Volunteer Force field battery based in Drogheda, by 1940 the battery was subordinate to the 4th Artillery Battalion in Western Command. In 1941 it was based in Longford, and was part of the 2nd Field Artillery Battalion. In 1941 the battery trained on the 4.5 inch howitzer. The 8th Field Artillery Battery has been subordinate to the 4th Field Artillery Regiment since the time of The Emergency. During The Emergency it was equipped with four 4.5 inch howitzers. Subsequently it was equipped with 18-pdrs, which were last fired by the battery in 1953 and which were replaced by eight 120mm Brandt mortars. In 1959 it was redesignated as the 8th PDF Heavy Mortar Battery; equipment issue was scaled back to six 120mm mortars in 1979. In 1998 the battery number ceased to exist as the PDF batteries were renumbered in accordance with a standardized scheme; the battery was renumbered as No. 1 Gun Battery, 4th Field Artillery Regiment, equipped with the RUAG 120mm mortar.

9th Field Artillery Battery (Templemore): Originally a Volunteer Force field battery based in Cork. The 9th Field Artillery Battery was subordinate to the 3rd Field Artillery Regiment during The Emergency, and in 1946 was disbanded along with the 3rd Regiment. On 1 October 1959, when the PDF and the FCA were merged, the Templemore Infantry Battalion FCA was redesignated as the 9th Field Artillery Battery and issued with eight 4.5 inch howitzers. Its first live firing practice was in August 1961, at the Glen of Imaal. In 1978 the 4.5 inch howitzers were replaced by the 25-pdr field gun. The battery had centers at Templemore, Knock, Curraganeen, Roscrea, Templetouhy, Athnid, and Castleiney. When the 3rd and 8th Regiments merged in October 2005 to form the 31 Reserve Artillery Regiment, the 9th Battery lost its name and historical identity and became HQ Battery, 31 Reserve Artillery Regiment.

An unusual view of a French 75 and crew in a field setting, probably during The Emergency. It appears that the gun is being manhandled into position. The gunner with the Enfield rifle slung over his shoulder seems particularly encumbered. (Colin Stone)

A 120mm Brandt mortar preparing to fire at the 6th Field Artillery Regiment camp at Stranahealy in 1994. The sergeant is wearing a brassard with the Curragh Command flash above his stripes. All detachment members are wearing active ear protectors, which was unusual for the period. The two water bottles that are visible are U.S. Army surplus water bottles. The layer is wearing a beret rather than a helmet because the turtle-back helmet style interfered with getting an eye close to the sights. (62 Reserve Artillery Regiment, via Capt Ian Stewart)

10th Field Artillery Battery (Dublin): Originally a Volunteer Force field battery based in Tralee. In 1940 the battery was attached to the 11th Infantry Battalion, Gormanston. The 10th Field Artillery battery has been subordinate to the 2nd Field Artillery Regiment since the time of The Emergency, when it was only one of two batteries to be equipped with the French 75mm gun. From 1959 through 2000, it was equipped with six 25-pdrs, and in 2000 equipment was upgraded to the 105mm L118 Light Gun. This battery was renamed as No. 1 Gun Battery, 2nd Field Artillery Regiment, in 1998.

11th Field Artillery Battery (Edenderry): Originally a Volunteer Force field battery based in Waterford. In 1941, as part of the 2nd Field Artillery Battalion, it was stationed at McKee Barracks, Dublin. This battery was subordinate to the 6th Field Artillery Regiment during The Emergency, when it was equipped with the 4.5 inch howitzer. Along with its parent 6th Field Artillery Regiment,

Artillery Day, 16 September 1989, Kildare. Members of the 11th Heavy Mortar Battery, 6th Field Artillery Regiment, preparing for competition with the 120mm Brandt AM50 mortar. The 11th Battery were the winners of the competition. The truck-mounted Giraffe radar can be seen in the background. (62 Reserve Artillery Regiment, via Capt Ian Stewart)

it was demobilized in 1945. As part of the 1959 reorganisation the Edenderry Infantry Battalion became the 11th Heavy Mortar Battery (FCA), subordinate to the reborn 6th Field Artillery Regiment (FCA). Between 1959 and 1977 it was equipped with eight 120mm Brandt mortars, and in 1979 the issue was scaled back to six 120mm mortars. The battery was disbanded and lost its historical identity when the 6th and 7th Regiments merged in October 2005 to form the 62 Reserve Artillery Regiment, and the resulting batteries of the new regiment were assigned new numbers (1, 2, and 3). Upon the merger of the 6th and 7th Regiments in October 2005 to form the 62 Reserve Artillery Regiment, the 11th Battery, along with the 56th and 6th Batteries of the 6th Regiment, formed the No. 2 and No. 3 Gun Batteries of the new 62 Regiment.

12th Field Artillery Battery (Birr): Originally a Volunteer Force field battery based in Clonmel. The 12th Field Artillery Battery was subordinate to the 3rd Field Artillery Regiment during The Emergency, and in 1946 was disbanded along with the 3rd Regiment. On 1 October 1959, when the PDF and the FCA were integrated, the Ely Infantry Battalion FCA was redesignated as the 12th Field Artillery Battery, subordinate to the reconstituted 3rd Field Artillery Regiment (FCA), and issued with eight 4.5 inch howitzers. In 1978 the battery's 4.5 inch howitzers were replaced by 120mm Brandt mortars and it was redesignated the 12th Heavy Mortar Battery. The battery maintained centers at Birr, Rathabbin, Lorrha, Borrisokane, Shinrone, and Clyduff. With the merger of the 3rd and 8th Regiments in October 2005 to form the 31 Reserve Artillery Regiment, the 12th Battery, along with A Coy, 14 Battalion, became 31 Regiment's No. 3 Gun Battery.

13th Field Artillery Battery (Tuam): Originally a Volunteer Force field battery based in Mullingar. The 13th Field Artillery battery was subordinate to the 5th Field Artillery Regiment during The Emergency, equipped with 4.5 inch howitzers, but along with its parent regiment, was disbanded in 1946. In 1959 a new 13th FCA Heavy Mortar Battery was formed, based on the Tuam Infantry Battalion FCA, and subordinate to the newly formed 5th Field Artillery Regiment. The battery was equipped with the 120mm Brandt mortar. When the 5th and 9th Regiments merged in October 2005 to form the 54 Reserve Artillery Regiment, the 13th Battery became the basis of No. 2 Gun Battery of 54 Reserve Artillery Regiment.

Members of the 13th Artillery Battery at McGee Barracks, Kildare posing along with two 4.5 inch howitzers in December 1945. (Colin Stone)

Paddy Delaney laying a 14th Battery 25-pdr, 1991 Camp. Delaney was a dedicated gunner, travelling from Kerry to Dublin to parade on Thursday nights. (Colin Stone)

14th Field Artillery Battery (Dublin): Originally a Volunteer Force field battery based in Longford. From The Emergency through 1977, the battery, designated the 41st Field Artillery Battery (later redesignated as the 14th Battery), was subordinate to the 2nd Field Artillery Regiment. During The Emergency its equipment consisted of 60-pdr guns, replaced by four 18-pdrs in 1945. Reportedly, four of the original nine 18-pdrs in the Irish inventory were in service with the 14th Battery. In 1949 the battery was issued with French 75s, which it retained until 1959 when they were replaced by 18-pdrs. Although equipped with 18-pdrs, nominally it was a 25-pdr unit. In 1974 it was equipped with the 4.5 inch howitzer. Ultimately, it was equipped with the 25-pdr. In October 2005 when the 6th and 7th Regiments merged to form the 62 Reserve Artillery Regiment, the 14th Battery, along with the 19th Battery, formed the basis of the HQ Battery and the No. 1 Gun Battery of the new 62 Regiment.

15th Field Artillery Battery: Originally a Volunteer Force field battery based in Letterkenny. In 1939 the battery was allotted to the northern coast artillery forts at Lenan and Dunree. During

The assembled members of the 14th Battery, taken at the Glen of Imaal in 1963. (Shay Stone via Colin Stone)

No. 1 Mortar Team, Glen of Imaal, in 1981. The team consists of Sgt Shay Stone (one of five brothers who served with 7th FAR), Sgt John Whelehen, and Cpl Joe McGuinness. It is interesting to note that the three individuals are wearing or holding a variety of head gear, including a non-regulation baseball cap, a 1944 pattern helmet, and a "soup bowl" or Tommy helmet. Sgt Stone's beard is also non-regulation, but despite all of the departure from regulations, the team, from 14th Bty, 7th FAR, managed to place No. 1 for the 1981 camp. The mortar has also had sandbags placed under the base plate to keep it from bedding in too deep, but this practice was later discontinued as the sandbags on occasion would be thrown out at high speed when the mortar was fired, potentially causing injury to the crew. (Shay Stone via Colin Stone)

The Emergency and until 1979 it was subordinate to the 4th Field Artillery Regiment, after which it became subordinate to the 9th Field Artillery Regiment. While subordinate to the 4th, initially it was equipped with four 4.5 inch howitzers, changed to six 25-pdr guns in 1959; in 1979, when it was resubordinated to the 9th Regiment, its equipment was once again changed, this time to six 120mm mortars. When the 5th and 9th Regiments merged in October 2005 to form the 54 Reserve Artillery Regiment, the 15th Battery became the HQ Battery of 54 Reserve Artillery Regiment.

16th Field Artillery Battery (Thurles): Originally a Volunteer Force field battery based in Ballinasloe. The 16th Field Artillery Battery was subordinate to the 3rd Field Artillery Regiment during The Emergency, and in 1946 was disbanded along with the 3rd Regiment. On 1 October 1959, when the PDF and the FCA were merged, the Thurles Infantry Battalion FCA was redesignated as the 16th Heavy Mortar Battery of the reconstituted 3rd Field Artillery Regiment (FCA) and issued with eight 120mm Brandt mortars. The battery had centers at Thurles, Gortnahoe, and Holycross. When the 3rd and 8th Regiments merged in October 2005 to form the 31 Reserve Artillery Regiment, the 16th Battery became No. 2 Gun Battery, 31 Reserve Artillery Regiment.

17th Field Artillery Battery: This number originally was assigned to a Volunteer Force battery. The 17th Field Artillery Battery was subordinate to the 7th Field Artillery Regiment during The Emergency. It was disbanded in 1945, and reappeared in 1979 subordinate to the 1st Field Artillery Regiment, equipped with six 120mm Brandt mortars. The battery number was retired in 1998

A 25-pdr of the 6th Field Artillery Regiment at the Glen of Imaal in 1985. Crew are wearing the old-style helmets and seem to be bereft of any ear protection. (62 Reserve Artillery Regiment, via Capt Ian Stewart)

when the batteries in the PDF regiments were simply renumbered as 1 and 2 within the regiments; No. 2 Gun Battery is the heir to the 17th Battery..

18th Field Artillery Battery: This number originally was assigned to a Volunteer Force battery. The 18th Field Artillery Battery was subordinate to the 7th Field Artillery Regiment during The Emergency; along with its parent regiment, it was disbanded in 1945 and reappeared in 1979 subordinate to the 2nd Field Artillery Regiment, equipped with six 25-pdr guns. The battery number was retired in 1998 when the batteries in the PDF regiments were simply renumbered as 1 and 2 within the regiments, and the 18th Battery was redesignated as No. 2 Gun Battery, 2nd Field Artillery Regiment.

19th Field Artillery Battery (Dublin): This number originally was assigned to a Volunteer Force battery. This battery was subordinate to the 2nd Field Artillery Regiment during The Emergency, equipped with the 4.5 inch howitzer. In 1959 the 19th FCA Heavy Mortar Battery was formed from the 42nd Field Artillery Battery, and equipped with eight 120mm Brandt mortars. In 1979 it was subordinated to the 7th Field Artillery Regiment, and the number of mortars was scaled back to six. In October 2005 when the 6th and 7th Regiments merged to form the 62 Reserve Artillery Regiment, the 14th Battery, along with the 19th battery, formed the basis of the HQ Battery and the No. 1 Gun Battery of the new 62 Regiment.

In order for the guns to be effective, vital direction and support are provided behind the scenes by command and signals personnel. This photograph, probably dating in the early 1950s, shows 2nd Lieutenant (later Commandant) Terry Lowen as the Gun Position Officer on the left, and Mick O'Beirne as the radio operator. (Shay Stone via Colin Stone)

20th Field Artillery Battery (Mullingar): This number originally was assigned to a Volunteer Force battery. This artillery battery was subordinate to the 4th Field Artillery Regiment from the Emergency until 1946, when it was demobilized. It was reactivated as part of the 4th Field Artillery Regiment in October 1959 as an FCA battery, equipped with six 18-pdr guns. In 1979 it was resubordinated to the 9th Field Artillery Regiment and equipped with six 120mm Brandt mortars. When the 5th and 9th Regiments merged in October 2005 to form the 54 Reserve Artillery Regiment, the 20th Battery became the basis of No. 3 Gun Battery of 54 Reserve Artillery Regiment.

21st Field Artillery Battery: This number originally was assigned to a Volunteer Force battery. The 21st Field Artillery Battery was formed during The Emergency; it was based in Clonmel in 1943-1945, and was equipped with four 4.5 inch howitzers. During The Emergency it was subordinate to the 1st Field Artillery Regiment, but with the drawdown of the Army in 1945-46, it was demobilized. In 1959 it was reconstituted, based on the independent 45th FCA Field Artillery Battery and renamed the 21st FCA Heavy Mortar Battery and was equipped with eight 120mm Brandt

The crew of No. 5 Mortar, 20th Field Artillery Battery (FCA), 9th Field Artillery Regiment, during camp at Glen of Imaal in 1981. From left to right: Gnr R. Ginnell, Gnr Jack Lynam, Gnr Carr, Sgt Gerry O'Connor, Gnr Stephen Corcoran. (Gerry O'Connor via Colin Stone)

A 120mm Brandt mortar at the moment of firing at Stranehealy woods in 1994. Only one crew member has covered his ears against the noise; the others may possibly be wearing ear plugs. (62 Reserve Artillery Regiment, via Capt Ian Stewart)

mortars. In 1979 it was resubordinated to the 8th Field Artillery Regiment and was equipped with six 120mm mortars. On 1 October 2005 it merged with the 2nd Field Artillery Battery to become the 1st Battery of 31 Reserve Field Artillery Regiment, equipped with four 25-pdrs.

22nd Field Artillery Battery: This battery had its origins as a Volunteer Force battery. Subordinate to Western Command during The Emergency; in 1979 it reappeared, subordinate to the 4th Field Artillery Regiment, equipped with six 105mm L118 Light Guns. Presumably, in 1998 this battery was renumbered as No. 2 Gun Battery of the 4th Field Artillery Regiment.

23rd through 28th Field Batteries: The numbers 23 through 28 were assigned to Volunteer Force Field Batteries, but never were incorporated into Field Artillery Regiments or otherwise became part of the active Army inventory.

41st Field Artillery Battery (Dublin): The lineage of the 41st Field Artillery Battery L.D.F. Can be traced to No. 7 Group, L.S.F in summer of 1940 at the old Cabra Garda Barracks. When the L.D.F was formed in autumn 1941, members of the Group were selected for transfer to the anti-aircraft artillery, and in June 1942 this group was transferred to Collins Barracks, Dublin, for field artillery training with the 2nd FA Battalion. In October 1942 or shortly thereafter, the 41st Field Artillery Battery L.D.F. was formally established, with Capt Jack Nolan as battery commander. The first battery shoots were fired on 7 August 1943. In March 1944 the 60-pdr guns were issued, and the 42nd Battery was also formed. Both batteries were assigned to defence of Dublin Bay, with the 41st Battery position in St. Anne's Park, Raheny. After The Emergency, the 60-pdrs were withdrawn from service and replaced by 18-pdrs for firing practice during the 1945 annual camp. During the 1947 and 1948 camps, the battery fired the 4.5 inch howitzer. In 1949 the battery was issued the French 75, which would remain on hand until 1959. On 8 October 1959 the battery moved from Collins Barracks to McKee Barracks. Upon formation of integrated regiments, the 41st Battery became the 14th Battery of 2nd FAR, and the French 75s were withdrawn and replaced by 18-pdrs. In 1962, following a short conversion course, the battery fired the 25-pdrs. On 28 April 1974 the battery fired its last 18-pdr round the 18-pdrs were replaced by the 4.5 inch howitzer. In 1979, the battery was subordinated to the 7th FAR.

42nd Field Artillery Battery (Dublin): The 42nd Battery was formed in March 1944, and during The Emergency it was equipped with 60-pdr guns. The battery was assigned to the defence of Dublin Bay, with a position in the Merrion area. After The Emergency, the 60-pdrs were withdrawn from service and replaced by 18-pdrs for firing practice during the 1945 annual camp.

Pulling a 60-pdr into position using drag ropes. This photograph was taken at Collins Barracks, Dublin, probably during The Emergency. (Shay Stone via Colin Stone)

During the 1947 and 1948 camps, the battery fired the 4.5 inch howitzer. In 1949 the battery was issued the French 75, which would remain on hand until 1959. On 8 October 1959 the battery moved from Collins Barracks to McKee Barracks. Upon formation of integrated regiments, the 42nd Battery became the 19th Battery of 2nd FAR, and the French 75s were withdrawn and replaced by 120mm Brandt mortars. On 28 June 1967, the 19th Battery, with four 120mm mortars, was airlifted to alternative firing positions during live firing practice. This was the first time that this type of exercise had been carried out in the Irish Army. In 1979, the battery was subordinated to the 7th FAR.

44th Field Artillery Battery (Cork): The 44th Field Artillery Battery was an independent FCA battery, formed in April 1950, possibly equipped with the 4.5 inch howitzer. In October 1959 the battery was redesignated as the 2nd Field Artillery Battery and became part of the 1st Field Artillery Regiment and was equipped with the 120mm mortar. In 1979 the 2nd Battery was resubordinated to the new 8th Field Artillery Regiment.

45th Field Artillery Battery (Cork): The 45th Field Artillery Battery was an independent FCA battery, formed in the early 1950s, and twinned with the 44th Field Artillery Battery. Equipment may have consisted of the 4.5 inch howitzer. In October 1959 the battery was

A 2 Star Gunner stands by while a round is loaded into the 120mm AM 50 mortar during 6th Field Artillery Regiment training in 1998. Ear protectors appear to be standard issue. (62 Reserve Artillery Regiment, via Capt Ian Stewart)

A 6th Field Artillery Regiment 25-pdr, Glen of Imaal, in 1976. To the left of the 25-pdr are two of the venerable 3.7 inch mountain howitzers. (62 Reserve Artillery Regiment, via Capt Ian Stewart)

A 3.7 inch mountain howitzer firing in the Glen of Imaal, 1976 Although the image is somewhat indistinct, the gun barrel is at the moment of full recoil, with the breech-block almost touching the ground. (62 Reserve Artillery Regiment, via Capt Ian Stewart)

redesignated as the 21st Heavy Mortar Battery and became part of the 1st Field Artillery Regiment and was equipped with the 120mm mortar. In 1979 the 21st Battery, along with the 2nd Battery, was resubordinated to the new 8th Field Artillery Regiment.

46th Field Artillery Battery (Mullingar): The author has uncovered only one passing reference to the existence of this battery; the reference infers that the battery was in existence sometime between 1948 and 1959, but provides no amplifying details as to location, subordination, or equipment.

1st Light Battery: This battery was formed in 1933 and was equipped with four Ordnance QF 3.7 inch mountain howitzers. One reference states that the battery also included the limited number of mortars available at that time, but neither the number or types of mortars were specified.[1]

1. *Adrian J. English, Irish Army Orders of Battle, 1923-2004, page 20.*

As a matter of conjecture, presumably these mortars would have been holdovers from the Civil War, likely limited to a handful or less of 3 inch Stokes trench mortars. Prior to allocation of 25-pdrs, the 6th Field Artillery Regiment fired the 3.7 inch howitzer as its artillery piece. The battery remained in existence until 1977, after which it was disbanded.

2nd through 7th Light Batteries: The 1st Light Battery was a Regular unit equipped with 3.7 inch howitzers, whereas the 2nd through 4th Light Batteries were Volunteer Forces batteries, equipped with, or at least trained to use, the 81mm Brandt mortar, and frequently were referred to as mortar batteries rather than light batteries. Home station of the 2nd Light Battery was Monaghan, of the 3rd was Kilkenny, and of the 4th was Castlebar. Although actual manning levels likely varied considerably from battery to battery, a training deployment to Coolmoney Camp in early 1936 by the 3rd Light Battery provides an idea of the approximate size of a light battery; the personnel present for training amounted to four officers, one acting sergeant, two corporals, and sixty five Volunteers. Evidence of the existence, other than notionally, of the 5th, 6th, and 7th Light Batteries is sparse. In June 1940 the Volunteer Forces light batteries were integrated into the infantry structure as mortar batteries.

Synopsis of Field Artillery Battery Lineage

Historic Battery Number	Current (2012) Successor Battery
1st FA Battery	No 1 Gun Battery, 1st Field Artillery Regiment
2nd FA Battery	No. 1 Gun Battery, 31 Reserve Artillery Regiment**
3rd FA Battery	Depot and School Battery
4th FA Battery	No successor battery
5th FA Battery	No. 2 and No. 3 Gun Batteries, 62 Reserve Artillery Regiment***
6th FA Battery	No. 2 and No. 3 Gun Batteries, 62 Reserve Artillery Regiment***
7th FA Battery	No. 1 Gun Battery, 54 Reserve Artillery Regiment
8th FA Battery	No. 1 Gun Battery, 4th Field Artillery Regiment
9th FA Battery	HQ Battery, 31 Reserve Artillery Regiment
10th FA Battery	No. 1 Gun Battery, 2nd Field Artillery Regiment
11th FA Battery	No. 2 and No. 3 Gun Batteries, 62 Reserve Artillery Regiment***
12th FA Battery	No. 3 Gun Battery, 31 Reserve Artillery Regiment
13th FA Battery	No. 2 Gun Battery, 54 Reserve Artillery Regiment
14th FA Battery	No. 1 Gun Battery, 62 Reserve Artillery Regiment
15th FA Battery	HQ Battery, 54 Reserve Artillery Regiment
16th FA Battery	No. 2 Gun Battery, 31 Reserve Artillery Regiment
17th FA Battery	No. 2 Gun Battery, 1st Field Artillery Regiment
18th FA Battery	No. 2 Gun Battery, 2nd Field Artillery Regiment
19th FA Battery	HQ Battery, 62 Reserve Artillery Regiment
20th FA Battery	No. 3 Gun Battery, 54 Reserve Artillery Regiment
21st FA Battery	No. 1 Gun Battery, 31 Reserve Artillery Regiment**
22nd FA Battery	No. 2 Gun Battery, 4th Field Artillery Regiment

** 2nd FA Battery and 21st FA Battery merged to form No. 1 Gun Battery, 31 Reserve Artillery Regiment

*** The assets of 5th, 6th and 11th FA Batteries were the basis of two successor batteries, No. 2 and No. 3 Batteries of 62 Reserve Artillery Regiment

Air Defence Artillery

The beginnings of the anti-aircraft component of the Artillery Corps have a somewhat amusing aspect, albeit typically in line with the military rationale for doing things. Although one normally thinks of the military establishment conducting staff studies to determine equipment requirements, manning levels, funding priorities, stationing arrangements, etc., prior to organizing a new unit, the Irish Army seems to have disregarded all of these, and other considerations, when it established the first anti-aircraft unit in the army.

In 1927 an army purchasing mission to Britain purchased four 3-inch 20 cwt medium anti-aircraft guns. Two of the guns had four-wheel carriages and were classified as mobile, while the other two guns were mounted on two-wheel carriages and classified as semi-mobile. In March 1928 the four guns were delivered to Islandbridge barracks (now Clancy Barracks), Dublin, and from October 1928 to January 1929 Lt Maurice P. McCarthy attended an anti-aircraft artillery course at Larkhill Artillery School in England. The Artillery Corps now had a trained anti-aircraft artillery officer and a battery's worth of guns but, for reasons unknown, the guns remained stored at Islandbridge and Lt. McCarthy was detailed to duties not connected with his anti-aircraft training.

Then, in mid-September 1931 Major (later Lt Gen) P.A. Mulcahy, stationed at Kildare, received a phone call from GHQ on a Friday afternoon, informing him that he was to form an anti-aircraft battery immediately with the four guns that were stored at Islandbridge. Mulcahy laughed at the order and hung up on the caller. The following morning he received another phone call from GHQ inquiring as to whether the battery was formed yet. He replied that indeed it was not, as he was already fifty men short of establishment, but was told that he had better get it done immediately. Mulcahy again took no action. Sunday passed, but on Monday morning the phone rang again, this time a friend of Mulcahy's at the other end. The friend asked: "Paddy, have you formed that battery yet?" to which Mulcahy replied "I haven't, and I have no intention of doing it." The friend then said "Well, I'm telling you, you'd better get it done today, and I'll be ringing you this afternoon and you'd better be able to tell me that the battery has been formed." The friend's admonition was taken more seriously than were the previous calls, and Mulcahy, after conferring with Comdt Paddy Maher, decided to appoint Lt McCarthy (who had taken the anti-aircraft course at Larkhill a few years previously) as the OC of the battery, and Lt Jimmy Dolan as his 2 I/C. McCarthy and Dolan were sent to Islandbridge to pick up the guns, while *ad hoc* arrangements were made

The 3 inch battery firing at Gormanston in 1943. By this time, although the anti-aircraft organization was still miniscule, the Artillery Corps had added much more modern equipment to the inventory, including the 40mm Bofors gun and the 3.7 inch heavy anti-aircraft gun. (Lt Col Edmond Fogarty)

The original anti-aircraft battery in 1932, with one of the battery's 3 inch guns. Although the Irish Army had taken delivery of the 3 inch guns in 1928, they remained in storage until 1931 when Ireland's first anti-aircraft battery was formed. (MA via Lt Col Edmond Fogarty)

to house the guns at some stables at Plunkett Barracks in Curragh Camp. A few men were detailed to Lt McCarthy's "command" and the battery was thus formed. That same afternoon Mulcahy's friend called again to check on the status of the battery, as he had promised, and Mulcahy was able to report with a more or less straight face that the battery had been formed, under the command of Lt McCarthy, and that they were stationed at Plunkett Barracks. All of this rather peremptory ordering about to form the battery was put into context the following day, when there was a Dail question scheduled for answer by the Minister of Defence. The question was: "Is it a fact that there are guns – anti-aircraft guns – rusting in Islandbridge Barracks and that there is no unit in the army to look after them," to which the Minister was able to state that "the answer to both questions is in the negative." Orders were published officially on Wednesday, 16 September, establishing the anti-aircraft battery at Kildare Barracks. The strength of the battery was two officers and forty-eight Other Ranks, but in fact, the guns still remained at Islandbridge until 24 January 1932, when they finally were transferred to Kildare. The grouping of the guns and personnel at one location did not, however, solve the problem of how to move the guns to a range for practice firing. The guns remained immobile due to a lack of suitable towing vehicles;[1] it was not until 8 July 1933 that the first tractor capable of towing the guns was delivered and was used to tow one of the guns to Tobertoby Camp in County Louth. A second tractor, provided on a loan basis from a civilian firm on 10 July, enabled the section to be moved on the same date. The first shoot of the battery was not until 18 July, 1933. A second tractor subsequently was obtained by the battery, and the battery became fully mobile in June 1934 when two more tractors arrived.[2] On 3 December 1934, the battery was officially designated as the 1st A.A. Battery.

A 3.7 inch heavy anti-aircraft gun during a firing drill, at or near maximum elevation. Its cruciform stabilizers are extended and leveled. During The Emergency the army fielded both mobile and fixed versions of the 3.7 inch AA gun. (Military Archives)

From this rather hurried birth in 1931, the anti-aircraft component of the Artillery Corps had to wait some eight years and an imminent military conflict in Europe before there were signs of any further significant expansion. In September 1938 the battery was moved to McKee Barracks, Dublin, due to the unstable situation in Europe, and was ordered to prepare for expansion. A year later, in September 1939, the anti-aircraft battery was expanded to a brigade with a strength of 1038 personnel, headquartered in Portobello (now Cathal Brugha) Barracks, consisting of two medium batteries, four light batteries, and a searchlight battery. Due to the availability of only four of the 40mm Bofors Mk I guns, there was only one gun for each light battery. The prospect for additional equipment was bleak, as in June 1940 the Irish government had begged Britain to complete delivery of equipment that had already been ordered, among which were sixteen 3.7 inch guns and sixty Bofors guns. Britain, reeling from the debacle at Dunkirk, had reprioritized her own equipment needs, leaving little hope that aid for Ireland would be forthcoming. The brigade organisation itself, for various reasons was short-lived, and in early 1940 was replaced by an anti-aircraft

1. *It would appear from information contained in Karl Martin's Irish Army Vehicles: Transport and Armour Since 1922, that the only vehicles in the Irish Army's transport inventory remotely capable of towing the 3-inch 20 cwt guns were the Leyland Subsidy Type A 3-ton lorries, but the numbers operationally available in 1931 are a matter of conjecture (in 1925, 20 of 33 were fit for service).*
2. *Although there are isolated references to tractors in connection with the Anti-aircraft Battery, the specific make, type, and model appear undocumented. It is possible that these tractors were powered by steam rather than by internal combustion engines.*

battalion. On 31 May 1941, in light of the fact that not all of the equipment required to fill out the battalion was going to materialize, the establishment was reduced to 653 personnel (36 officers, 125 NCOs, and 492 Gunners).

It should be noted that the Defence Force Regulation 49 of 22 October 1934 that established the tactical and territorial organisation of the Defence Forces called for four Volunteer Force anti-aircraft batteries (2nd Medium and 2nd through 4th Light), and also designated the Regiment of Pearse as an anti-aircraft artillery unit on mobilization.

The Air Defence Command was established in autumn 1939. It is interesting to note that at the time the Command asked the *Garda Siochana* to cooperate by reporting the presence or movement of aircraft over certain of the larger towns and to telephone these reports to the Air Defence Command Headquarters at Dublin Castle. By 1939, the Artillery Corps mustered an AA artillery battalion at Army Headquarters level, consisting of four 3 inch, two QF 3.7 inch, and four Bofors L/60 anti-aircraft guns.

In 1939 the air defence of Dublin consisted of the Air Defence Command, an anti-aircraft battalion with three medium and one light batteries (the medium batteries were equipped with the 3 inch 20 cwt and 3.7 inch guns, and the light battery with the 40mm Bofors guns), and an understrength searchlight battalion, as well as fighter interceptors based at Baldonnel. In March 1939 fourteen gun and searchlight positions were selected for the defence of Dublin, generally concentrated in an arc on the eastern edge of the city, stretching generally from Howth in the north to Dalkey in the south, a defense clearly oriented against aircraft approaching from the general direction of Dublin Bay. There is some discrepancy in accounts as to whether guns were actually emplaced beginning in September 1939, or if allocation of the available guns took place between 31 May and 30 November 1940. Ultimately, two of the 3.7inch heavy guns were emplaced at Clontarf (Vernon Avenue) behind the tram sheds in static positions[3], and four were at Ringsend in the mobile configuration; only the two guns at Clontarf had all of their necessary fire control instruments. At least two of the Ringsend guns were in Ringsend Park, near Pigeon House Road, and the other two may have been in a position at Sean Moore Park. In 2009, residents local to the Ringsend area suggested that one possible reason for positioning the guns there was that during The Emergency, Ringsend Park was used as a storage area for fuel arriving from the UK, and for coal as well, so the guns would be able to protect the vital fuel supplies. The old 3-inch guns were located at Stillorgan (also referred to by some sources as Booterstown or Trimbleston Estate), near the present Montrose Hotel, and in Ballyfermot. Two of the 40mm Bofors Light Guns were at Baldonnel and two at the Hibernian Schools (now St. Mary's Hospital in Phoenix Park), to be moved to Collinstown in March 1941. Ammunition availability for the guns on 22 October 1940, was given as 1090 rounds for the 3.7 inch guns, 8900 rounds for the 3 inch guns, and only 1200 rounds for the 40mm Bofors guns. A battery of 12-pdr guns was also allocated for ground defence in conjunction with the anti-aircraft guns, one each of these guns being located at Collinstown, Hibernian Schools, North Bull and Sandycove. Searchlights and sound locator sections[4] were deployed at Howth Summit, North Bull Wall, South Bull Wall, Blackrock Park, Sandycove, and Dalkey. In 1942, delivery of a further twelve 3.7 inch heavy guns allowed three more gun positions to be established at Brownsbarn (near Baldonnel), Brackenstown (north of Dublin Airport), and Elm Park (near Booterstown).

This author has found one reference to an anti-aircraft defence battery stationed at Shannon Airport during The Emergency. Although this certainly would be logical, there seems to be no documentation available regarding this battery, nor to the type of equipment it may have had.

The first shots fired in anger by Irish artillery since the end of the Civil War occurred on 23 April 1941, when the 3 inch 20 cwt guns at Trimblestown fired six rounds against a German aircraft. The

3. *The actual position of the guns likely was at what is now Clontarf Park, about a hundred yards to the northwest of what were the tram sheds in 1940, and what is now a bus maintenance facility. The area is known locally as "The Sheds," in reference to the former tram sheds.*
4. *The searchlights and sound locators were taken over from the Engineers in 1939.*

3.7 inch guns at Ringsend AA station engaged another lone German aircraft later that same night; no hits were registered in either engagement. The following day, 24 April, four rounds were fired from the Ballyfermot position (near Phoenix Park) at an aircraft approaching from the north of the gun position. The most devastating bombing of the war occurred on 31 May, when an estimated twenty *Luftwaffe* aircraft bombed a number of areas in Dublin, causing widespread death and destruction. At 2348 hours on 30 May, unauthorized aircraft were heard east of Carnsore, County Wexford, moving north; and Air Raid Message "Yellow" was given to the guns and searchlights of the Dublin AA defences. At 2058 hours the aircraft had reached Wicklow Head and were still proceeding north, and Air Raid Message "Red" was given. At 0035 hours on 31 May the Clontarf gun position fired four 3.7 inch rounds, and at 0128 hours Ringsend fired four 3.7 inch rounds and Contarf fired another four 3.7 inch rounds. At 0130 hours the Stillorgan 3 inch position fired three rounds, while at the same time the Collinstown position opened fire with two 40mm Bofors guns and four .303-inch Hotchkiss machineguns, which fired 54 rounds and 248 rounds respectively. The 3 inch guns at Ballyfermot fired four rounds at 0131 hours. The Collinstown position engaged a twin-engined aircraft at 0145 hours, firing 58 rounds of 40mm and 490 rounds of .303-inch ammunition. The air raid alert ended at 0258 hours. Despite the anti-aircraft fire, the Germans managed to drop a number of bombs, beginning at about 0130 hours. The bombs fell in a mainly working class district of the city, including the areas of Summerhill Parade, North Richmond Street, North Circular Road, Newcomen Bridge, Phoenix Park, the Dublin Zoo, Rutland Place, and the North Strand, which was the hardest hit of all by a bomb that exploded at 2:05 a.m. The raid killed 28 people, injured 90, and destroyed approximately 300 houses, leaving some 400 people homeless. Although there are some theories to the contrary, the bombing most likely was the result of a navigational error and a mistaken target; the *Luftwaffe* had targeted Belfast on previous occasions and may simply have bombed Dublin by mistake. None of the attacking aircraft were downed despite the most sustained and intense anti-aircraft fire delivered during The Emergency.

A 3.7 inch mobile anti-aircraft position at Ringsend Park, Dublin, in 1941. The rangefinder in the foreground is a Barr & Stroud UB 7 that in turn is connected to a predictor that provided the bearing, elevation, and fuze setting to the gun. This is one of the guns that saw action on the night of 31 May, 1941 against German bombers that mistakenly attacked Dublin. (Hanley collection, MA, via Colin Stone)

Ringsend Park, September 2009. This is the same location as the photograph of the 3.7 inch mobile gun position that was taken in 1941. Although trees now obscure the buildings shown in the 1941 picture, close inspection reveals them to be the same buildings. (Roy Kinsella)

Following The Emergency, the Proposed Peace Establishment of December 1946 that regulated the reduction of army strength and organisation, the anti-aircraft component was reduced to a single training regiment with two heavy and one light batteries, with the heavy batteries equipped with the sixteen 3.7 inch guns and the light battery with the six 40mm Bofors guns then on hand. In 1949 the anti-aircraft inventory gained another eighteen Bofors L/60 AA guns, bringing the total to 24 Bofors guns. In the early 1950s, six British No. 3 Mk 7 radars were also procured and were used in conjunction with the 3.7 inch guns. The 3 inch 20 cwt guns had been retired in 1945/46, followed by the remaining 3.7 inch guns whose last shoot was in 1959, leaving

An L60 in a static display position. All four wheels are off the ground, and the gun is supported by the leveling jacks under the extended stabilizing outriggers. The L/60 guns represented the light air defence element of the Irish Artillery Corps during The Emergency. (MA, Artillery Box)

An undated photograph of the power operated 40mm Bofors being shown to visiting military attachès. From left to right are Lt. Col. A. Dalton, Col. A.F. Polk (U.S. Military Attachè), Maj Gen P.A. Mulcahy (Chief of Staff), Col. P. Curran (Director of Artillery), and Brig. R.N. Thickrisse (British Military Attachè). (Mullingar Barracks Museum via Colin Stone)

the Bofors guns as the sole type of anti-aircraft weapon in the inventory. The batteries were then reorganised and reequipped exclusively with the Bofors guns.

The integration of the FCA with the PDF in October 1959 resulted in a significant change in the anti-aircraft establishment. Two FCA batteries were formed, the 2nd in Dublin and the 3rd in Limerick; both batteries were equipped with 40mm Bofors L/60 guns and had a large FCA establishment and a small regular training cadre. In 1960/61 a 4th Anti-aircraft Battery was formed provisionally for deployment to the Congo in response to strafing of Irish positions there, but it was stood down after the Fouga Magisters that represented the air threat had been shot down. In 1979, with the addition of the 4th Anti-aircraft Battery (Reserve) to the organisational establishment, there were four six-gun batteries (one Regular and three Reserve), deployed throughout Ireland (Dublin, Kildare, Cork, and Limerick).

In 1980 Irish air defence capabilities were significantly upgraded by the introduction of the Swedish RBS 70 short-range man-portable surface-to-air missile system. The acronym RBS stands for Robotsystem, and the system falls into the category generally referred to by yet another acronym, MANPADS (for Man Portable Air Defence System). The RBS 70 was chosen by Ireland due to its low cost and low maintenance requirements, as well as for its effectiveness. The RBS 70 systems in the Irish

An RBS 70 at Gormanston shortly after launch. The RBS 70 is the only missile system in the Irish air defence inventory. The initial lot of six systems purchased in 1980 was supplemented by an additional purchase in 2008. (Lt Col Edmond Fogarty)

inventory are assigned to the 1st Air Defence Battery, Kildare.

In order to provide automated target acquisition and tracking for the RBS 70 systems, in 1986 the Swedish Ericsson Giraffe PSS-70 (also referred to as the Giraffe 40) radar system, mounted on a MAN 20:280 DFAEG 6 x 6 truck was delivered. The Giraffe radar can also be used to supply target information for the Bofors L/70 guns as well as for the RBS 70 missile system. The Giraffe 40 target acquisition and tracking radar is described as an Agile Multi Beam (AMB) three-dimensional search radar designed to operate as a central search/acquisition radar and control center for short-range anti-aircraft defence. The Giraffe radar can be employed in all types of terrain and has the capability to detect rocket, artillery, and mortar projectiles within a range of 40 km and can track such projectiles within a range of 20 km. When linked to nine RBS 70 firing posts (each 4 km apart) the tracking system can protect an area of up to 175 km^2. In 2008 the single truck-mounted Giraffe 40 was replaced by a number of improved Giraffe G 50 radars mounted on Swedish Bv 206 tracked vehicles, and an unspecified number of additional RBS 70 firing units were also procured second user at that time.

The air defence organisation has remained stable since the 1980s, with the 1st Air Defence Regiment Hq Battery and 1st Air Defence Battery in Kildare (representing the Regular component of the air defence establishment), and the 2nd through 4th Reserve batteries. Likewise, the equipment inventory, although upgraded, has remained substantially the same in terms of numbers and general equipment types.

In 1959 the Bofors L/60 guns were joined by two more modern 40mm L/70 guns obtained as experimental or test guns. The guns which were power operated were centrally controlled by a fire director which permitted both guns to be fired together at a single target using a reflex sighting system. It was not until much later, in 2001, that Ireland acquired thirty Bofors EL/70 guns along with eight Signaal Flycatcher radar target acquisition and fire control systems previously in service with the Royal Netherlands Air Force. The L/70 guns were a marked

The RBS 70 launch unit as a static display at Cathal Brugha Barracks, 5 May 2011. The Hagglunds BV 206 in the background carries the Giraffe Mk 4 radar associated with the RBS 70. (Colin Stone)

A Hagglunds BV 206 with its Giraffe Mk 4 radar deployed on 30 July 2009. Note the deployed stabilizing outriggers on the trailing unit. An RBS 70 air defense missile fitted to its tripod is in the foreground. (Colin Stone)

The RBS 70 air defense missile system, with the deployed Giraffe radar mounted on a BV 206 in the background. (Colin Stone)

A Bofors L/70 firing at Gormanston in 2005. The L/70 is a significantly improved version of the WW II Bofors L/60 that served as Ireland's primary light air defence weapon system from 1939 to 2001. (Irish Defence Forces)

The fully extended Giraffe radar on the 6 x 6 MAN truck. The outriggers have been lowered to stabilize and level the platform during operation. (Internet)

improvement over the long serving Bofors L/60 guns that by this time were obsolete, excellent as they may have been in their time.

The Flycatcher radar system is a self contained all-weather integrated search radar and fire control system. It is also fitted with an advanced electronic countermeasures suite, making it a reliable fire control system. The Flycatcher can track up to sixteen targets simultaneously within a range of 20 km, providing accurate targeting for its accompanying battery of six EL70 weapons systems, each of which can be deployed up to 2 km from the radar. Flycatcher is a radar system primarily designed to control the fire of air defence guns, but is capable of being upgraded to control the fire

A Flycatcher air defence radar on exercise. Eight Flycatcher systems were acquired from the Royal Netherlands Air Force in 2001, providing a substantial increase in target detection and tracking capabilities. (Colin Stone)

A Signaal Flycatcher radar van hitched to a Mercedes 1117A, parked on Cemetery Hill. The Flycatcher was purchased from Holland in 2001 to support the Bofors L/70 anti-aircraft guns purchased at the same time. (Colin Stone)

of short-range air defence missiles as well. The radar is simple to operate, with a single operator controlling the system from target detection until successful engagement. The system is equipped with modern electronic warfare countermeasures and is capable of operating in a hostile electronic environment. The Flycatcher was originally purchased in 1987 for the Dutch Army, and upgrades to the Dutch Flycatcher systems were completed in 1995. The Flycatcher systems purchased by Ireland were among those that were upgraded.

The Anti-Aircraft Brigades, Battalions, Regiments and Batteries : In September 1939 the seminal anti-aircraft battery was expanded to a brigade, headquartered in Portobello Barracks (now Cathal Brugha Barracks), numbering 1038 personnel and consisting of two medium batteries, four light batteries, and a searchlight battery. Due to the availability of only four of the 40mm Bofors Mk I guns, there was only one gun for each light battery. The brigade organisation was short-lived, however, and in early 1940 was replaced by an anti-aircraft battalion.

The Mercedes 614D artificer's van that was included as part of Dutch Bofors deal, but which is infrequently used. The vehicle still retains its original Dutch colors and markings. (Colin Stone)

A pair of 3 inch 20 cwt guns of the anti-aircraft battalion firing at Gormanston in 1943. Prior to being expanded as a battalion, all of Ireland's anti-aircraft assets were grouped as a single battery, consisting of the 3 inch 20 cwt guns based at Kildare. (Lt Col Edmond Fogarty)

The Anti-Aircraft Battalion: The Anti-Aircraft Battalion was formed in early 1940, retaining the same equipment as its predecessor, the Anti-Aircraft Brigade, and maintaining a similar manning level. When it became apparent that additional air defence equipment was not forthcoming, the decision was made to reduce the manning establishment to levels commensurate with the equipment actually on hand, and on 31 May, 1941, battalion strength was reduced to 36 officers, 125 NCOs, and 492 Gunners, for a total of 653 personnel (a strength reduction of some 385 personnel). The battalion, configured with three medium batteries and one light battery, remained throughout The Emergency, with appropriate increases in manning as new equipment arrived. The component batteries, whose equipment types and numbers varied, appear not to have been numbered, but instead may have been referred to simply by the location at which they were stationed (Ringsend, etc.). With the drawdown of the Army following The Emergency, the AA Battalion was dissolved in 1946, and was replaced by four "Fixed AA Regimental Cadres," two in Dublin and two in Cork.

Anti-Aircraft Training Regiment (Kildare): On 16 June 1953, the fixed regimental cadres were replaced by an Anti-Aircraft Training Regiment with headquarters in Kildare, with a flexible battery organisation. This organisation remained until October 1959, when it was replaced by an organisation that included FCA elements.

Operating a range finder during anti-aircraft live-fire training at Gormanston in 1943. Although range finding and fire control equipment was basic during the early part of The Emergency, generally it reflected contemporary technological capabilities. (Lt Col Edmond Fogarty)

1st Air Defence Artillery Regiment (Kildare): With the integration of the FCA in October 1959, the anti-Aircraft Training Regiment was redesignated as the 1st Anti-aircraft Artillery Regiment with headquarters, along with the HQ Battery and the 1st Battery, in Kildare. The 2nd Battery was located in Dublin, and the 3rd Battery was based in Limerick; both of these batteries were FCA batteries. Following the "disintegration" of the Army in September 1979, the 1st AA Regiment remained integrated, with one regular (1st Battery) and three FCA batteries (2nd, 3rd, and the 4th Battery which had been organised in 1979). In the 1984/1985 timeframe, anti-aircraft units were redesignated as air defence units, and the regiment thus was renamed the 1st Air Defence Regiment. The Air Defence Regiment is the only Irish artillery unit to have deployed to the European mainland to conduct training and live firing practices alongside one of its European counterparts, the Swedish Air Defence Regiment Lv6, and the air defence element of the Swedish amphibious forces.

1st Air Defence Battery (Kildare): The 1st AA Artillery Battery was formed on 14 September 1931, and officially established by orders published on 16 September 1931. The battery consisted of four 3 inch 20 cwt guns and was commanded by Lt Maurice McCarthy. Its first shoot took place almost two years later, on 18 July, 1933. Between the time of its formation in September 1931 and December 1934 the battery was known simply as The Anti-Aircraft Battery; on 3 December 1934, it was formalized as the 1st Anti-Aircraft Battery. The battery made its first public appearance in the St. Patrick's Day Parade of 1935. In 1936, in conjunction with the Air Corps, it began firing practices at Gormanston Camp, County Meath. During The Emergency it was one of the batteries of the Anti-aircraft Brigade, and its successor, the Anti-aircraft Battalion. In the post-war period it was equipped with 40mm Bofors L/60 guns. In 1984/1985, the designation of the battery was changed to the 1st Air Defence Battery; the 1st Air Defence Battery has remained the premier air defence battery in the Artillery Corps, and retains its status as a Regular battery. It is equipped with six RBS 70 short range air defence missile systems, as well as with six 40mm Bofors L/70 guns.

2nd Air Defence Battery (Dublin): The precursor of the 2nd Air Defence Battery was the 2nd AA Artillery Battery (FCA), which came into existence on 1 October 1959, with the reorganisation of the PDF. At that time, ten officers selected from FCA infantry units in and around Dublin were collected together at Griffith Barracks and transferred to the new unit, under the command of Lt P. Holahan. In 1984/1985 the battery's designation was changed from anti-aircraft artillery to air defence. The battery, located at Cathal Brugha Barracks, Dublin, was equipped with the 40mm Bofors L/60 gun until 2001, when equipment was upgraded to six 40mm Bofors L/70 guns. The battery is a Reserve battery, and is subordinate to the 1st Air Defence Regiment.

A Bofors EL/70 in a firing position at Gormanston. Although the system is located on a level concrete hardstand, the stabilizers are lowered to optimize system performance. (Lt Col Edmond Fogarty)

3rd Air Defence Battery (Limerick): Along with the 2nd Air Defence Battery, the 3rd Air Defence Battery came into existence on 1 October 1959, as the 3rd AA Artillery Battery (FCA), with the reorganisation of the PDF. The first shoot of the 3rd AA Battery took place in the Glen of Imaal in July 1960. The first time the guns of 3rd AA battery fired at an aerial target was at Gormanston in 1978. In 1984/1985 the battery's designation was changed from anti-aircraft artillery to air defence. In 2001 the battery's 40mm Bofors L/60 guns were replaced by the more modern and capable Bofors L/70 guns. The battery is a Reserve battery and is subordinate to the 1st Air Defence Regiment.

4th Air Defence Battery (Cobh): The original 4th Air Defence Battery was formed in 1960/61 for deployment to the Congo, but was stood down when the air threat there ceased to be a factor after the Fouga Magisters which had been used to strafe Irish positions had been shot down. The current 4th Air Defence Battery was formed in 1979 and traces its lineage back to the coast artillery defence unit based at Fort Mitchel, on Spike Island in Cork Harbour. On 1 September 1979, when the CDA unit was disbanded, it was re-roled as an anti-aircraft unit, and equipped with six 40mm Bofors L/60 guns and subordinated to the 1st Air Defence Regiment. On 21 March 1985, renamed the 4th Air Defence Battery, the battery was moved from Spike Island to Belmont Military Post in Cobh and more recently to the Naval Base on Haulbowline. The battery is a Reserve battery subordinate to the 1st Air Defence Regiment, and is equipped with six 40mm Bofors L/70 guns.

The Flycatcher radar controls and directs the fire of the Bofors EL/70 guns. Eight of these systems were acquired from The Netherlands in 2001, along with the purchase of thirty of the EL/70 guns, thus providing the Artillery Corps with an integrated capability. (Department of Defence)

Coast Artillery

As part of the Anglo-Irish Treaty of 6 December 1921, Britain was allowed to retain control over the so-called "Treaty Ports" at Cork, Bantry Bay, and Lough Swilly. Integral to the agreement was control of the coast artillery installations in those locations, amounting to seven separate forts (Forts Camden, Carlisle, Westmoreland and Templebreedy in the Cork Harbour area, Bere Haven in Bantry Bay, and Forts Dunree and Lenan on Lough Swilly on the west coast). In early 1938, Britain and Ireland began negotiations to resolve the issue of ultimate control over the forts; despite serious resistance by Winston Churchill, the British generally were well disposed to signing the forts over to the Irish Government, and on 25 April 1938, a new agreement was signed in London that provided for the forts to be turned over to Ireland.

Just prior to takeover of the forts in 1938, the Irish Army had formed its own coast artillery service with headquarters on Spike Island. Once the Irish authorities had assumed ownership of the forts, three of the Cork Harbour forts were renamed: Fort Camden became Fort Meagher, Fort Carlisle became Fort Davis, and Fort Westmoreland became Fort Mitchel. Templebreedy's name was unchanged, as were the names of the forts at Bere Haven, Dunree, and Lenan. The forts at Lough Swilly remained in British hands somewhat longer than those at Cork; it was not until 3 October that those two forts were formally taken over on behalf of the Irish Government. Thus, by October, the service consisted of six coast batteries, three of which were based at Cork, together with the headquarters on Spike Island. The total number of guns amounted to at least seventeen (five 9.2 inch Mk X, and eight Mk VII six inch, and four Mk IV 4.7 inch), plus an undetermined number of 12-pdrs (possibly as many as twenty-one).[5] The 9.2 inch guns were, and likely will remain, the heaviest artillery pieces ever in the Irish artillery inventory. At the time of turnover, the two 9.2 inch guns at Fort Templebreedy were found to be defective; one was removed and replaced by the 9.2 inch gun from Fort Davis, while the other was repaired, resulting in a net loss of one of

One of the two 6 inch coast defence guns at Fort Davis, 21 July 1987. The guns at Davis had been inactive for over twenty years when this photograph was taken; although still retaining its dazzle paint scheme, the gun is showing signs of rust and deterioration. To the left of the photo are Fort Meagher (ex-Camden) in Crosshaven, and beyond that, what appears to be an island is Curraghbinny, a finger of land that juts into the harbour and is now a national forest. Spike Island is barely visible in the right background. (Denis McCarthy)

5. *Duggan, page 232, mentions the figure of twenty-one 12-pdrs; the number of 12-pdrs is difficult to determine precisely. Four each can be accounted for at Fort Davis and at Dun Laoghaire; there may have been eight to twelve at Bere Island. The possibility of 12-pdrs at Duncannon Fort bears further research. There also are isolated references to two torpedo tubes, of unspecified size, at both Fort Camden and Fort Davis, but this author has been unable to confirm this as fact.*

The 6 inch gun position at Newhaven Fort, Sussex, England. The 6 inch gun in this emplacement was obtained from Ireland in 1988. In Ireland it had originally been installed at Duncannon Fort, Co. Wexford, Ireland. (Ed Tyhurst, via Philip Baldock)

the 9.2 inch guns. The coast defence artillery inventory was augmented by two Mk VII six inch guns installed at Fort Shannon in 1942, bringing the total number of operational guns to at least nineteen. Ammunition supplied with the guns, however, was not considered abundant, as there were only 170 rounds for each of the 9.2 inch guns, and 275 rounds for each 6 inch gun. This impacted on the ability to conduct practice shoots. In addition to the fixed coast artillery guns, reportedly there were also a number of 18-pdr guns and 4.5 inch howitzers at the forts that were not included in the turnover, but were purchased separately. Six 18-pdrs were at the Cork forts, and two 4.5 inch howitzers were located at Bere Haven and another four at Lough Swilly; these guns were used to equip field batteries rather than being retained in the coast defence role. In 1942, six 60-pdr guns were acquired from Britain, and although they were mobile artillery, some sources assert that they were assigned to the coast artillery, but evidence on this point is not conclusive.

An interesting sidelight concerns the 6 inch gun presently situated at Newhaven Fort in Sussex, England, acquired from Ireland in 1982. The gun, a BL 6 inch Mk VI on a Mk II mounting, originally was located at Duncannon Fort, Co. Wexford, protecting the approach to Waterford. A plate on the gun indicates that the gun was from Duncannon Fort, Waterford, Eire, No. 2261; the plate also indicates that the trunnion height of the gun was 55 ft 0 in above M.S.L. (mean sea level) at Duncannon. The gun in question, however, is reported to have gone to England from Spike Island, so it would appear that at some point the gun had been transferred from Duncannon Fort to Spike Island (possibly in 1922, as the armament from Duncannon Fort was removed by the British prior to turning over the fort to the Irish authorities). Correspondence between Newhaven Fort authorities and the Directorate of Artillery, McKee Barracks, Dublin, in 1981 also makes reference to four 12-pdr 12 cwt guns stored in Depot Ordnance. Three of the guns were described as Elswick pattern guns manufactured by Wolseley Motors in 1917, and a fourth was marked "Manufacturer VS&M 1/1902." One of the 12-pdrs was acquired by Newhaven Fort and is mounted there; that particular gun is reported as having come from Clancy Barracks "on the Liffey." The correspondence reveals that also stored in Depot Ordnance at the time were two 4.7 inch guns destined for Fort Duncannon, but never mounted there; one of these guns presumably is now on display at Dunree.

At their peak strength, the British coast defence garrisons in Ireland had amounted to some 800 personnel. Even after their drawdown following World War I, peacetime garrisons had been maintained, the installations remained operational, and the guns were maintained in an operationally ready posture. When the forts were turned over to the Irish, there remained a need for training on

A 12-pdr at Newhaven Fort, Sussex, England. This 12-pdr was obtained from Ireland around 1988, along with a 6 inch gun, and was restored prior to display. (Ed Tyhurst, via Philip Baldock)

the guns. The British had agreed that a cadre of three officers, eighteen NCOs, and six other ranks should remain in Ireland to assist in the training of the coast artillery personnel, as the operation and maintenance of the fixed coast artillery guns differed markedly from field artillery equipment that Irish Artillery Corps personnel were familiar with. Coast artillery gunnery procedures had to take into account not only the significantly different gun characteristics, but considerations such as rapidly changing atmospheric conditions, firing at moving targets at long ranges, targets whose aspect ratio could change quickly, changes in sea level due to the rise and fall of the tide, and barrel wear, among other factors. The British instructors did in fact remain for an unspecified time and worked along with the new Irish coast artillery gunners in a relationship of mutual respect and professional cooperation. The instructors apparently moved from fort to fort to provide instruction. By all accounts, the British were good teachers and the Irish were apt pupils.

Selection of gunners for the coast artillery began in 1938. Prospective gunners had to be a minimum height of 5'8". Recruits were initially selected by being stood up against a chalk line marked on a wall; those who reached or surpassed the 5'8" mark could go to the Artillery Corps; in one intake of 1000 recruits in January 1938, 360 were tall enough to be coastal gunners. The height requirement was necessary to be able to manhandle the ammunition for the large guns.

Coast artillery gunnery training was carried out throughout the period of The Emergency, on both the 6-inch and the 9.2 inch guns. Gunnery practice at Templebreedy was carried out by the 9.2 inch guns by firing at a target float some 12 feet long. Reportedly, during the first shoot by Irish gunners, they destroyed the target float at a range of five miles, attesting to the quality of the training and the skill of the gunners. Prior to firing the 9.2-inch guns, residents in the area were warned to open their windows in an attempt to avoid broken glass from the concussion of the guns. All of the 6-inch shoots in Cork Harbour were fired from Fort Davis, day and night during the summer months. Night shoots were aided by powerful searchlights that illuminated the targets. A launch, making about nine knots, would tow three target floats behind on a thousand yards of cable. Firing would begin when the launch would begin to wind in the cable, adding the winching speed to the speed of the launch. The story is oft repeated that the 6-inch guns were so accurate, and the crews so skilled, that an experienced crew could, and often did when they wanted to, cut the tow rope. With all due respect to the gunners, such an assertion seems a bit fanciful especially when one reads accounts of the target launch itself being inadvertently hit or suffering near misses. Training at Bere Island was similar to that at the Cork forts, with training culminating in annual firing

A 6 inch gun at Fort Davis, Cork Harbour. The photograph shows the state of deterioration of the shield and of the gun tube itself due both to the exposure to the elements and lack of any type of maintenance or preservation efforts. (Sean Murphy)

practices in July and August. Likewise, the first shoot of the Lough Swilly guns took place in the summer of 1940. Records indicate that while other forts could equal or exceed the accuracy of the 9.2-inch guns at Fort Lenan, over the years their speed in action seldom could be beaten.

With the passing of The Emergency, the coast defence installations soon felt the effects of the Army's downsizing in 1945-1946. The forts assumed cadre status, and practice or demonstration shoots became infrequent, if not nonexistent. By 1950, the 9.2 inch guns, as well as some of the still serviceable 6 inch guns, were in a state of care and maintenance. The two obsolete 4.7 inch guns were still serviceable, and the 12-pdrs were used only for saluting purposes. The guns of Dunree were silent from 1945 until 1964, when a 6 inch practice shoot was carried out; the final shoot at Dunree took place in July 1967. In 1969, Fort Davis saw a shoot of its 6 inch guns. By 1977, the coast artillery was reduced to a cadre in Cork Harbour; the guns, maintained in heavy preservation, had been turned over to the 4th Field Artillery Regiment in Mullingar.

The coast artillery, having served well and faithfully during The Emergency, soon saw its role eclipsed by the advent of technologically advanced weaponry that rendered coast defence obsolete worldwide. In about 1983, with no coast artillery guns remaining in an operational status, a combination of safety, storage, and financial considerations dictated the disposal of the remaining stores of coast artillery ammunition. Accordingly, the task was passed to the Irish Naval Service which dumped the now useless ammunition at sea. The ship involved in the disposal was the LE *Setanta* (formerly the Irish Lights tender *Isolda*).

Coast Artillery Forts

Fort Meagher (formerly Fort Camden) is the only purely Palmerstonian fort in the Cork defences, the other forts being adaptations and extensions of existing structures. There is evidence of fortifications on the site dating back to the mid-16th century – in the late 1600s the fort was known as James' Battery and consisted of two blockhouses and eight guns – although construction on the land front of Fort Camden was not begun until 1862. The fort is positioned on Rams Head covering the western entrance to the harbor, with Fort Carlisle opposite, on the eastern side of the harbor entrance. In 1938, when the fort was turned over by the British, it was bereft of armament, its operational six inch guns having been removed in 1904. The Second Earl of Camden (John Pratt), who was Lord Lieutenant of Ireland from 13 March 1795 to 14 June 1798, was the original namesake of the fort. When the fort was turned over to the Irish government in 1938, the name

Four 12-pdr salute guns under protective tarpaulins, East Pier, Dun Laoghaire, in a 1987 or 1988 photograph. As of early 2010 these same four guns remained in place as a ceremonial battery. A similar battery, also cocooned, is maintained at Fort Davis in Cork Harbour. (Denis McCarthy)

promptly was changed to Fort Meagher, commemorating Thomas Francis Meagher who was notable from a number of aspects. He was born in Waterford City, in County Waterford in 1823, was an Irish nationalist and revolutionary (for which he was transported to Tasmania, but from which he escaped), and is credited with designing the Irish tricolor flag. Upon his escape from Tasmania he emigrated to New York, and when the American Civil War broke out, he joined the Union Army, rising to the rank of Brigadier General, forming and leading the Irish Brigade.[6] Following the Civil War, he served as acting governor of the Montana Territory, and in 1887, under mysterious circumstances, he fell off a steamboat in the Missouri river and drowned. The fort was relegated to caretaker status with the end of The Emergency in 1946. The Department of Defence handed over Fort Meagher to the Cork City Council in May 1989, which had plans to develop the fort as a tourist attraction. However, the upper levels of the fort served for at least some time as a storage area for Southern Command and as a training area for both the Army and Naval reserves. On 22 July 2010 the Cork City Council handed the fort overt to the Rescue Camden Committee, and restoration work to develop the site as a tourist facility was begun by a group of local volunteers whose goal it is to clean, restore and develop the fortress to its former glory, creating a tourism and heritage centre. Fort Camden is internationally recognized as one of the world's finest remaining examples of a classical coast artillery fort.

Fort Davis (formerly Fort Carlisle) had been built on the eastern tip of Cork Harbour in 1797, and between 1860 and 1868 the fort's present form began to take shape. In 1880 casemates were excavated in the face of the cliffs and equipped with rifled muzzle-loading guns; by 1901 armament consisted of two 9.2 inch, two 6 inch, and four 12-pdr guns. The fort was taken over from the British on 11 July 1938, at the same time as the Spike Island installation. At the time of the turnover, armament at Fort Carlisle consisted of two 6 inch Mk VII guns, and one 9.2 inch Mk X gun in the South Battery, which soon was removed to replace a defective 9.2 inch gun at Fort Templebreedy. The 6 inch gun emplacements were unprotected positions devoid of overhead cover, and thus exposed to enemy fire. All 6 inch shoots in Cork Harbour were fired from Fort Davis. The fort was originally named after Frederick Howard, 5th Earl of Carlisle, who was Lord Lieutenant of

6. *The Union Army's Irish Brigade, as its name implies, was made up mainly of Irish immigrants to America. The brigade originally consisted of the 63rd, 69th, and 88th New York Infantry regiments, and was later joined by the 116th Pennsylvania and 28th Massachusetts Infantry regiments. The brigade's core regiment, the 69th, gained notoriety prior to the Civil War when its colonel, Michael Corcoran, refused an order to parade the regiment for the Prince of Wales during his visit to New York City. The 69th Regiment earned the sobriquet "The Fighting 69th". Its war cry was faugh a ballagh, a corruption of the Gaelic fág an bealach, or "clear the way." The 69th fought in both World Wars, and the 1st Battalion, 69th Infantry served in Iraq in 2004-2005.*

Ireland from 1780 to 1782. In 1938, the name of the fort was changed to Fort Davis, in honor of Thomas Osborne Davis who was born in Mallow, County Cork in October 1814, and later became a chief organiser and leader of the Young Ireland movement. Although a Protestant, he preached unity between Catholics and Protestants, and dedicated his life to Irish nationalism. His accomplishments for the Irish cause were cut short by his early death from tuberculosis in September 1845. Almost exactly eight years after it had been taken over from the British, Fort Davis was closed on 12 July, 1946, but its guns were put under care and brought to life each year by the gunners of Fort Mitchel. The final shoot may have occurred in 1969. A saluting battery of four 12-pdr guns is still maintained at the fort for ceremonial occasions.

Fort Mitchel (formerly Fort Westmoreland) Spike Island, in Cork Harbour, at one time a smuggler's haunt, was bought by the British in 1790 and fortified with a battery of 24-pdr guns in anticipation of a possible French invasion. Fort Mitchel occupied the high ground on Spike Island and

One of the Fort Davis 6 inch guns in 1987, with a concrete blockhouse behind it. Fort Davis was on the eastern side of Cork Harbour, facing Fort Meagher on the western side of the harbour; Fort Meagher, however, was devoid of guns during The Emergency. (Denis McCarthy)

The ammunition storage gallery at the Ruperts Tower Battery, Fort Davis, July 1987. Early in the 20th Century there were also two 6 inch guns at the North Battery in Fort Carlisle (now Davis), but by 1938 only the two 6 inch guns at the Rupert's Tower battery remained, along with a 9.2 inch gun which was soon removed to Templebreedy. (Denis McCarthy)

One of the 12-pdr salute guns at Fort Davis in Cork Harbour, seen without its protective tarpaulin, revealing the high state of maintenance that allows them to continue to be used for ceremonial saluting purposes. (Gabriel Whitson)

Three of the four 12-pdr salute guns while firing a ceremonial salute. Note the ear protectors under the berets being worn by all personnel. (Gabriel Whitson)

One of the 12-pdr salute guns at Fort Davis uncovered, being prepared to fire a salute. As shown in other photographs, these guns are normally covered to protect them from deterioration. (Irish Defence Forces)

was a regular six bastioned work surrounded by a ditch. In 1847, no longer under threat of French invasion, Spike Island became a penal settlement until 1883, when it again reverted to military control until 1920-1921, when some 500 Irish prisoners were held during the War of Independence. On 11 July 1938 the Irish government finally gained control of the coast artillery forts at Cork and their armament. The Spike Island defences were taken over, amidst some ceremony, at 1810 hours; shortly thereafter the *Taoiseach* (Prime Minister), Eamon de Valera and a number of other dignitaries arrived, and at 2000 hours the Irish tricolor was hoisted over Cork Harbour. Armament turned over to the Irish Government in 1938 consisted of two Mk VII six inch guns, in exposed positions on the ramparts. In 1942 the guns were relocated to new underground positions, protected by concrete casemates built by the Corps of Engineers. In 1943 further improvements were made with the construction of a new DEL (Defence Electric Light) engine room under the ramparts. Although the fort's guns never had occasion to be fired in defence of the harbour, they were manned round the clock from 1 September 1939 until the end of the war. The fort was originally named after John Fane, 10th Earl of Westmoreland, who served as Lord Lieutenant of Ireland from 1789 to 1794, and who visited the installation in 1791. The 1938 name change memorialized John Mitchel, who was born in Camnish, County Londonderry, in November 1815, and became a leading member of both Young Ireland and the Irish Confederation. Mitchel has an intimate connection with Spike Island; having been convicted of treason in 1848, he was held there prior to being transported to Tasmania, from which he escaped in 1853. He made his way to the United States, where he supported the Confederate cause; upon his return to Ireland, he was elected as a Member of Parliament for County Tipperary in 1875, but died suddenly on 20 March, 1875. On 21 March 1985 Fort Westmoreland was handed over to the Department of Justice; the 6 inch guns remain in place, and four 12-pdrs are still maintained on the island as salute guns for ceremonial occasions. On 12 July 2010 Spike Island was handed over by the Department of Justice and Law Reform to the Cork City Council; some effort at restoration may be attempted.

Fort Templebreedy, about two miles south of Crosshaven, is situated on the western approach to Cork Harbour, outside the harbor proper. Construction of the fort was begun in 1904 and was completed in 1909; armament consisted of two 9.2 inch, two 6 inch, and four 12-pdr guns. When the fort was turned over by the British in July 1938, the armament consisted of two powerful 9.2 inch Mk X guns on disappearing carriages. During the turnover it was discovered that the tube of one of Mk X guns was cracked and that the other was suffering a constriction of the barrel near the muzzle, referred to as steel choke. The cracked gun was replaced by the 9.2 inch gun taken from the South Battery at Fort Davis and the steel choke of other gun was repaired by the British by October 1938, when the repaired piece was mounted and calibrated. The name Templebreedy

The original exposed positions of the two 6-inch guns at Spike Island, before they were relocated to protected positions. (Colin Stone)

One of the 6-inch Mk VII guns at Fort Mitchel, on Spike Island in Cork Harbour, taken in 2007. The gun is in an excellent state of preservation. The bucket below the breechblock held water for swabbing the bore between rounds. Note the gunners' jackets hanging in the background. (Gabriel Whitson)

No. 1 6 inch gun in its revetted position, somewhat overgrown with gorse bushes, on Spike Island. This gun is in obvious need of restoration. (Gabriel Whitson)

A very nice aerial view of Spike Island and Fort Mitchel, which covers a large portion of the island's surface. (Gabriel Whitson)

A 9.2 inch coastal gun being prepared for firing at Templebreedy. The four 9.2 inch guns were the heaviest artillery pieces ever to serve in the Irish Army; two of the guns were located at Templebreedy protecting Cork Harbour, and two were at Fort Dunree on Lough Swilly. Note to the left of the crew the large bucket and the pole mounted brush used to swab the barrel after firing. (photograph courtesy Irish Examiner)

is a corruption of the Gaelic Teampall Bride, or possibly for St. Brigid's Church. Templebreedy was closed in April 1946, but the guns were fired once more in the fifties. Subsequently, the guns were melted as scrap by Irish Steel. Fort Templebreedy is owned by the Department of Defence.

Bere Island is situated about a mile south of the Castleownbere and covers the approaches to Bantry Bay. As early as 811 the Vikings established a base at Lonehort Harbour on Bere Island. Following the abortive attack by a French invasion fleet in 1796, in 1798 four Martello towers were erected on Bere Island, at Lonehart, Rerrin, Cloonaghlin West, and Ardagh Hill, with a gun battery built near each tower. The stone for these installations came from Cork and from the south of England, in addition to locally quarried stone. Work on the current fortifications began in 1898, and by 1903 batteries had been installed at Lonehort, Ardaragh, and Rerrin on the eastern end of the island and at Derrycreevin, Reenduff, and Ardnakinna on the western end. By 1914 Rerrin had a 9.2 inch gun, and Lonehort had two 6 inch and one 9.2 inch guns; Derrycreevin had two 6 inch guns, Reenduff had two 4.7 inch guns, and Ardnakinna and Ardagh each had two 12-pdrs. When the Bere Island defences were turned over on 29 September 1938, the remaining armament consisted of two 6-inch guns at Lonehort, plus two 4.7 inch guns at Reenduff, and two 12-pdrs and two 4.5 inch howitzers at other locations on the island. During the war, training was carried out on a daily basis, and, in addition, the coast artillery provided boarding parties for ships seeking

One of Templebreedy's 9.2 inch guns firing, sometime during the 1940s. These guns were the southernmost coast defence battery in the Cork area and controlled the entrance to Cork Harbour. (Irish Press)

One of the Mk VII 6 inch guns at Lonehort Battery on Bere Island. Although there is a plan to develop the site as a tourist attraction, as of mid-2011 there has been little or no progress toward that end, although a herd of wild mountain goats has been introduced to trim back the overgrowth. The guns themselves have been deteriorating since the mid-1940s and are in a sad state of disrepair. (Colin Stone)

The two derelict 6 inch guns at Lonehort battery. The brick structure in the foreground was the guard house, and the concrete building on the hillock is what remains of the battery command post. (Colin Stone)

One of the long since abandoned gun positions at the western end of Bere Island. The emplacements at the western end of the island were armed with 6 inch, 4.7 inch, and 12-pdr guns. (Colin Stone)

temporary shelter in Bere Haven, and search parties to locate the wrecks of foreign military aircraft that had strayed and had crashed in the West Cork mountains. Bere Island closed down in 1947, but as of 1973 the guns were still there under heavy care and preservation. As of 2012 only the two 6-inch guns at Lonehort remain in place, but are now in rather deplorable condition as they are in open positions, and the ravages of time and weather have taken an obvious toll. In November 2007 Bere Island Projects Group, Ltd., commissioned a consultancy firm to examine Lonehort Battery and draw up a five-year plan that could lead to the opening of the battery as a new tourist site for the island. The study recommended the opening of the fort on a three phased basis, with the first phase opening in June 2010. The study felt that the 6 inch guns could be restored to a standard of a similar gun at Newhaven Fort in Sussex, England.

Fort Lenan in Inishowen, County Donegal, is the northernmost of the Lough Swilly forts. Work was begun on the fort in 1901, and the following year it was operational. It had three concrete gun positions sited on a cliff facing seaward; it mounted 9.2 inch heavy guns, served by magazines beneath them. Along with Fort Dunree, Fort Lenan was taken over from the British on 3 October 1938. Armament at the time consisted of two 9.2 inch Mk X coast defence guns. The Lenan guns were known for their speed in action, if not for their accuracy on all occasions. In July and August 1940 the gunners from Lenan were moved to Fort Dunree, leaving only a maintenance party in place. Sometime, believed to be in the 1960s, the guns at Lenan were cut up and disposed of as scrap. Fort Lenan is now owned by the Donegal County Council and is in a state of disrepair.

Fort Dunree: In 1798 a small fort situated on a rocky promontory on the western side of the Inishowen Peninsula, County Donegal, mounting two 6 inch guns, was erected to guard against the possibility of attack by a French invasion fleet. In the late 19th century the fort was modernized and enlarged with the building of the "Top fort" on Dunree Hill. During World War I the fort stood guard over Admiral Lord Jellicoe's fleet anchored in Lough Swilly prior to engaging the German fleet in the Battle of Jutland. The fort, with its two 6-inch guns, was taken over by the Irish government on 3 October 1938; the last British troops left Dunree on 10 October. The 6 inch guns at Dunree were manufactured at Woolwich Arsenal in 1912. The guns were in exposed rather than casemated positions. Fort Dunree now hosts the Fort Dunree Heritage Museum, which opened in 1986, and which has a large collection of coast defence weapons on display, including the 6-inch guns originally mounted at Dunree as well as the 6 inch guns formerly located at Fort Shannon. The fort was used for military training until the 1980s, and the Department of Defence officially withdrew from Dunree in 1990. As with the fort at Lenan, Fort Dunree is now owned by the Donegal County Council, and the Museum is operated by a private corporation.

One of the two 9.2 inch guns in Lough Swilly's Fort Lenan in 1940. Note the two colours of paint on the barrel. (Colin Stone)

Fort Shannon was an Irish fort from start to finish. During The Emergency, a determination was made that the small examination port of Cappa, on the Clare side of the Shannon Estuary would need artillery support. The idea of protecting the estuary was further strengthened by British urging, who suggested that an installation be constructed at Ardmore Point some fifteen miles from the sea, as the flying boat terminus at Foynes, which was a vital strategic link for the Allies, lay further upstream. The British, in fact, agreed to provide two 6 inch Mk VII guns with which to arm the fort. This led to design and construction of a coast defence artillery installation to be named Fort Shannon, near Tarbert, on the Kerry side of the estuary. Accordingly, a five acre plot of land was purchased from a local landowner, and arrangements also were made with other landowners on whose lands local defence posts were to be built. This was the first major seaward defence to be planned and built by the Irish themselves. The project involved a major rock blasting program to provide overhead cover for the guns, the magazines, and for associated facilities, many of which were connected by tunnels. Two engineer companies, plus an engineer section, and a company from the Construction Corps were assigned to the construction tasks. A network of radio, telephone, and visual communications was developed between nearby posts, installations, and commands. Construction of the fort proper began in August 1942, and in November 1942, a mere four months after work had begun, the fort had been completed. The completed

One of the 6 inch gun positions at Fort Dunree, with a commanding view of Lough Swilly. The guns at Dunree are in a much better state of preservation than any of the other 6 inch guns in exposed positions in Ireland. (Colin Stone)

Cpl Gerry O'Connor and Gunner 2 Star Eric Craig posing next to one of Fort Dunree's 6 inch guns in 1978. (Gerry O'Connor via Colin Stone)

installation, manned by troops who had been brought in from the Cork Harbour forts and from Kildare Barracks, consisted of a two-gun battery of Mk VI 6 inch guns, a platoon of machineguns, and a searchlight. The location of the fort was ideal, as it commanded a stretch of the navigable channel that restricted maneuver of potential enemy vessels; in addition, the guns blended in well with the terrain and were difficult to spot from the estuary. Fort Shannon was closed on 31 May 1946, and in 1989 the two 6 inch guns were removed and sent to Fort Dunree, where they remain on display.

The gun emplacement of No. 2 gun at Fort Shannon, as seen from the river. This emplacement was on the left flank, and mounted a Mk VII gun on a PIII naval mounting. The gun mounted in this emplacement, serial number 1927, was manufactured in 1903 and issued in 1904, and last calibrated in 1937 before delivery to Ireland. (Courtesy of Donard deCogan)

One of the ex-Fort Shannon 6 inch guns that was moved to Fort Dunree in 1986. The Shannon guns were not part of the original inventory of coast artillery pieces turned over by the British with the Treaty Ports in 1938, but were supplied by Britain in 1942 for protection of the Shannon Estuary. (David Magee, Fort Dunree Museum)

A 12-pdr at Fort Dunree. A hallmark of the Dunree guns is the red oxide paint used to provide a modicum of protection from the deteriorating effects of the sea air and the elements in general. (Colin Stone)

Artillery Equipment

The great preponderance of artillery used by the Irish Army is of British origin. All of Ireland's field, anti-tank, and coastal artillery have been supplied by Britain. Air defence artillery has been largely supplied by Britain as well, even though the Bofors L60 guns acquired from Britain were a Swedish design, and the later L70 guns were not supplied by Britain but by The Netherlands. The RBS 70 air defence missile system also is a Swedish product. The one type of artillery which historically has not been of British origin is the mortar; although a number of mortars were supplied by Britain during The Emergency, the numbers pale in comparison to those supplied by France, South Africa, and most recently, Switzerland.

Artillery nomenclature

Tube artillery falls into three categories, described as guns, howitzers, and mortars. Various methods or standards are used to describe the differences between the types, but the most common method of understanding the difference relates to the angle of fire of the weapon. A gun fires in a flat or relatively flat trajectory (anti-tank guns are a classic example); a howitzer fires in a curved trajectory or an arc; mortars fire at an extremely high trajectory. Guns and howitzers generally have rifled bores (although some modern guns have smooth bores), whereas mortars almost always have smooth bores. Mortars are the simplest of artillery pieces, and in oversimplified terms, are nothing more than a piece of well-engineered pipe with a closed breech which has a firing pin in it which serves to activate the mortar bomb, which is simply dropped down the muzzle end of the barrel. There are some exceptions, as some very heavy mortars are breech-loaded weapons, but by and large, mortars are simple, inexpensive weapons organic to infantry elements, providing fire support down to company level.

Field Artillery

Canon dc 75 modèle 1897 (French 75)

The Canon de 75 modèle 1897 was the iconic French field gun, adopted in March 1898, and saw widespread use during World War I. It was referred to colloquially in French as the *soixante-quinze* (literally, "sixty-fifteen," which is the French manner of expressing the number seventy-five).

Part of French 75 battery and crews during a training exercise in the Glen of Imaal in 1975. Among the crew members, presumably of the gun in the foreground, are Sgt. Billy McMahon, Billy McCormack, Cpl Tony Freeney, and O'Reilly (Christian name not provided in the original caption). (Shay Stone via Colin Stone)

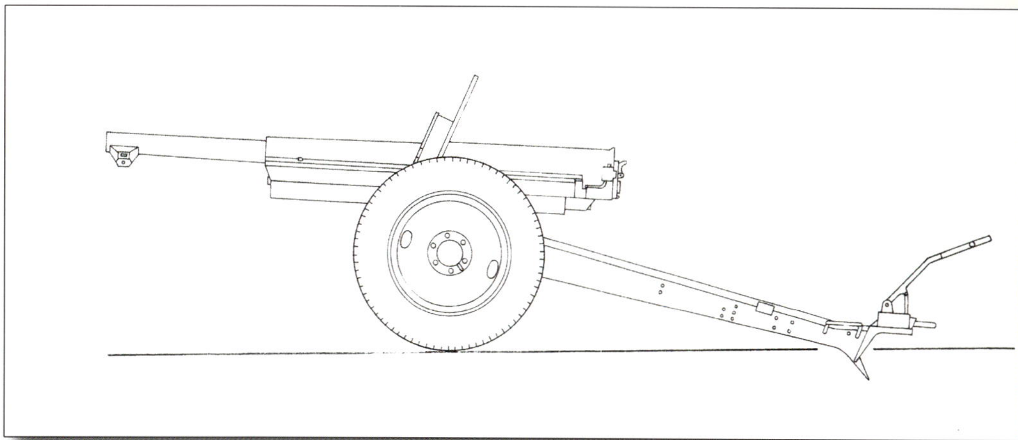

Breech detail of the restored French 75mm gun at Dromod, County Leitrim. The gun is finished in the grey that was in use for Irish artillery pieces for a time, beginning in 1926. (Philip Bedford)

One of the French 75s supplied by the British. The prominent "ears" below the muzzle are a distinctive feature of the French gun. The two-hole steel wheels and the axle hub mark this as a Martin Parry carriage. (Sean Murphy)

An Irish French 75 from the rear, showing the gun's characteristic breech and the single trail. The Land Rovers Series III LWB and the trucks in the background of this picture taken at Depot Kildare suggest the picture was taken in the late 1970s. (Sean Murphy)

The 75mm gun at Aberdeen Proving Ground. The difference between the Irish 75mm guns with the single trail and the U.S. 75mm guns with the split trail carriage is easily observable in this photograph. (Ralph Riccio)

A French 75 at Infantry Square, Curragh Camp, in October 2009. The gun has been retrofitted with early-style wood spoke wheels. Although the gun at first glance appears to be in decent condition, in actuality it has been only cosmetically refinished and has several troubling rust spots, the wheels themselves are quite rotted in some places, and the paint colours are far from authentic. (Sean Reilly, via Ian Stewart)

The French 75 introduced the revolutionary hydro-pneumatic long recoil system that kept the gun's wheels and trail stable during firing, avoiding the bounce and rearward displacement characteristic of guns prior to the 75. This enabled the gun to sustain a high rate of fire on target (up to 15 rounds per minute). Some 21,000 were produced between 1897 and about 1940. The French 75 was used principally by France in WW I, but by WW II was in service in Britain, Poland, the U.S., and Ireland as well, with captured examples used by the Germans. A modified form of the French 75 was the main armament for the WW II U.S. *Sherman* tank. Eight French 75s were supplied to Ireland from British stocks in December 1941.[1] They equipped the 10th and 14th Batteries of the 2nd Field Artillery Regiment and remained in service until 1968.

1. *One source claims that the U.S. supplied twelve 75mm guns in 1941, but photographs of the 75mm guns in Irish service show them to be the French pattern 75s rather than the modified US versions.*

A pair of French 75s dug in and camouflaged, possibly during The Emergency, being inspected by Col P.J. Hally, Adjutant-General. The camouflage is fairly effective, and digging the guns in lowers their profile significantly. (Shay Stone via Colin Stone)

Designation	Canon de 75 modèle 1897, or Matériel de 75mm Mle 1897.
Manufacturer	French state arsenals (Bourges, Chatellerault, Puteaux, St. Etienne, Tarbes)
Dates in Irish service	1941 – 1968
Number in Irish service	8
Caliber	75mm
Crew	6
Elevation	-9° 14' to + 45° 30'
Traverse	30° left and right
Weight	3404 lbs (1544 kg)
Length	NA
Width	NA
Height	NA
Carriage	Single trail
Barrel length	102 inches (2588mm)
Breech	Revolving block
Recuperator	Hydro-pneumatic
Ammunition types	HE, shrapnel
Shell weight	11.7 lbs (5.3 kg) HE; 16 lbs (7.24 kg) shrapnel
Muzzle velocity	1640 fps (500 m/s)
Rate of fire	15 rounds per minute
Range	9350 yards (8550 meters) HE; 7437 yards (6800 meters) shrapnel

Loading a high explosive round into a French 75. The gun fired fixed ammunition and had a high rate of fire, normally 15 rounds per minute, but could reach up to 30 rounds per minute for a short period with a highly experienced crew. (Colin Stone)

Ordnance QF 18 Pounder

The 18-pdr was introduced in British service in 1904; its design resulted from a 1902 Equipment Committee created to guide development of a quick-firing gun to replace the existing BL 15-pounder field gun. The carriage recoil system for the 18-pdr was developed by Vickers, and the gun itself was developed by Armstrong. During World War I, the 18-pdr was used on all fronts. During the period between the two World Wars, it was in a reserve status, although some were converted to the later very successful 25-pdr. The original Mk I gun had a wire-wound barrel that was stronger, lighter, and cheaper to manufacture than a built-up barrel. However, beginning in 1906, the Mk II replaced the Mk I; the Mk II changed from the wire-wound barrel to a gun sleeve with an inserted inner liner, and also had an armoured box-shaped oil reservoir on the front end of the

Front view of a trailer, or limber, for an 18-pdr. These were towed between the tower and the gun itself, and carried ammunition as well as various tools and accoutrements specific to the 18-pdr. (Defence Forces manual, via Sean Murphy)

Carriage, 18-pdr, Mk II, rear view. This is an unmodified 18-pdr, with the original pole trail and wood spoke wheels. (Defence Forces manual, via Sean Murphy)

Rear view of the artillery trailer for the 18-pdr. The trailer carried 38 rounds for the gun. While in transit, the flap below the open ammunition rack was folded up and closed to protect the rounds from dirt, dust, mud, and inclement weather. (Defence Forces manual, via Sean Murphy)

Front view of an 18-pdr, Mk IIPA, taken from a manual on the gun. The Martin Parry carriage was fitted with two-hole steel wheels and pneumatic tires replacing the original wood spoke wheels. (Defence Forces manual, via Sean Murphy)

Carriage, 18-pdr, Mk IIPA, rear view, with the early single pole trail, modernized with a Martin Parry conversion kit that replaced a number of axle components to allow fitting with pneumatic tires. (Defence Forces manual, via Sean Murphy)

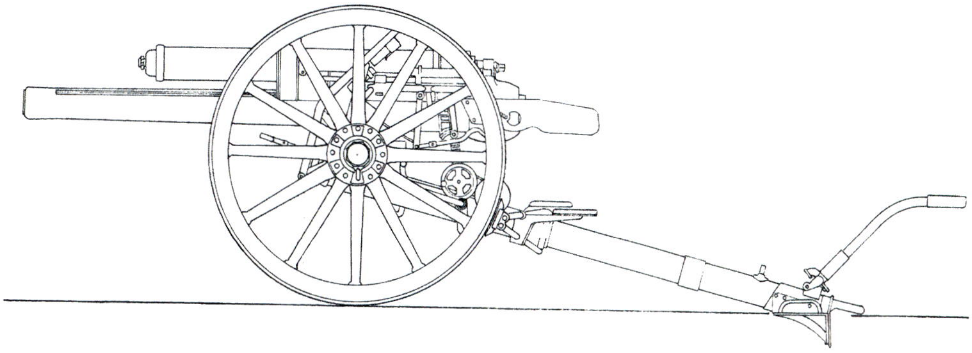

Ordnance QF 18-pdr, Mark I

recuperator to maintain the oil supply in the recuperator and to extend spring life. One of the distinguishing features of these early 18-pdrs was the recuperator mounted above the barrel, that was significantly shorter than the barrel itself. Both types were mounted on the Mk I carriage that featured a single pole trail that suited it to towing by teams of horses. In 1922, the British transferred nine 18-pdrs to the National Army; of these, four were the early Mk I version, and five were the improved Mk II. The 18-pdr thus was the first artillery type used by the fledgling National Army, and indeed was the only type of field artillery used during the Civil War in 1922. Aside from one gun captured by the Republicans and subsequently destroyed rather than allow its recapture by the National Army, all of the 18-pdrs were used by government forces against the Republicans. Between 1925 and 1941 Ireland continued to obtain small lots of 18-pdrs of various Marks, mainly the Mk IV, from Britain, and by 1941 the total number in the inventory stood at 33 guns. The 18-pdrs originally were fitted with wooden wheels befitting their World War I origin; although the wooden wheels were suitable for being towed at relatively slow speed by horses, with the increasing advent of motorization and trucks used as prime movers, upgrading to pneumatic tires was imperative. Consequently, some carriages were converted to the Martin Parry Mark II P[2] configuration with pneumatic tires, but still retaining the pole trail; others were obtained with the Mark IV carriage with a box trail or the Mark V carriage with a split trail and pneumatic tires. These configurations allowed for high-speed towing by motorized prime movers. For a period (around 1940), guns that had yet to be upgraded were carried portee-style in truck beds for transport. The 18-pdrs began to be replaced in 1949 by the Ordnance QF 25-pdr field gun, also obtained from Britain. Although

2. *Although the Martin Parry carriages were obtained from Britain, the carriages were built by the Martin Parry Company of York, Pennsylvania, which had filled an order for 2,500 of the carriages for Britain. Martin Parry also supplied a similar carriage for the US 75mm field pieces.*

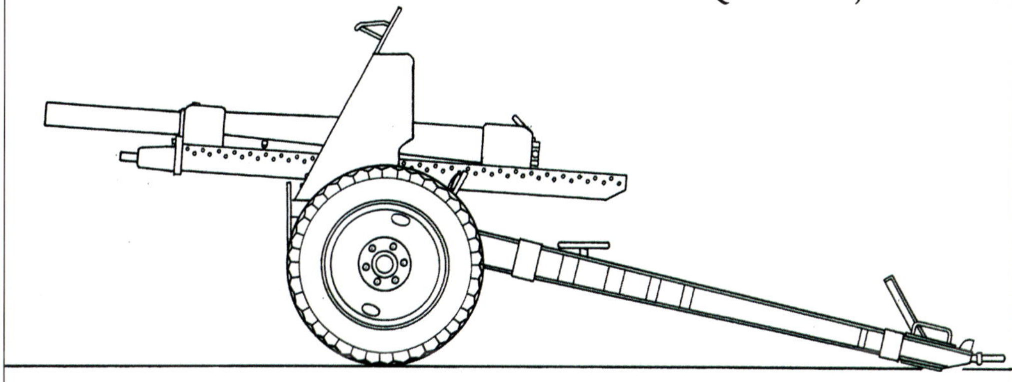

Ordnance QF 18-pdr, Mark IV

An 18-pdr of the 6th Field Artillery Regiment at a shoot in the Glen of Imaal in 1960. The breech of the gun is open, ready to accept another round. Five rounds remain in the limber, while two rounds are in the hands of gunners to the left of the gun. (62 Reserve Artillery Regiment, via Capt Ian Stewart)

some batteries were nominally 25-pdr batteries, they remained in fact equipped with the 18-pdr until the 1970s. The last 18-pdr round was fired in the Glen of Imaal by the 14th Battery, 2nd Field Artillery Regiment, on 28 April 1974, marking the end of the service life of this iconic gun that served the Irish Army for more than fifty years.

Designation	Ordnance QF 18-pdr (various Marks, including I, II, IV and V)
Manufacturer	Vickers-Armstrong
Dates in Irish service	1922 – 1974
Number in Irish service	33 – 37 (sources differ as to total number in inventory)
Caliber	3.3 inch (83.8mm)
Crew	6
Elevation	-5° to +16°
Traverse	4.5° left and right
Weight	2831 lbs (1284 kg)
Length	198 inches (5029mm)
Width	75 inches (1905mm)
Height	NA
Carriage	Pole trail (Mk I and II); box trail (Mk IV carriage); split trail (Mk V carriage)
Barrel length	92.62 inches (2462mm)
Breech	Screw (Mk 1); hydro-pneumatic (Mk 2)
Recuperator	Springs
Ammunition types	HE, AP, smoke, canister, gas, star
Shell weight	HE 18.5 lbs (8.39 kg); including case 23 lbs (10.43 kg)
Muzzle velocity	1615 fps (492 m/s)
Rate of fire	20 rounds per minute
Range	6525 yards (5966 meters)

A Mk IIPA 18-pdr upgraded with pneumatic tires, shown during a training exercise. This gun is most probably one of the four original Mk II guns that formed the nucleus of the Artillery Corps in 1923. (Colin Stone)

The restored 18-pdr at Dromod in 2006. The split trails of the Mk V carriage are folded for towing. As with many historical military equipment items in private collections, the standard of restoration and maintenance is often superior to that found in similar pieces maintained by government entities. (Philip Bedford)

The Cavan & Leitrim Railway Museum's restored 1919 18-pdr in 2006, seen from the rear. The restoration is marked by an extreme attention to detail. (Philip Bedford)

An 18-pdr at Connolly Square, the Curragh, 9 April 2010. This well preserved example maintained by the Artillery Corps is one of the few examples of restored or preserved guns in the Artillery Corps collection that is displayed in the open; others, such as the 3.7 inch mountain howitzer, the 4.5 inch howitzer, and the 17-pdr, are kept indoors and are not exposed to the elements. (Colin Stone)

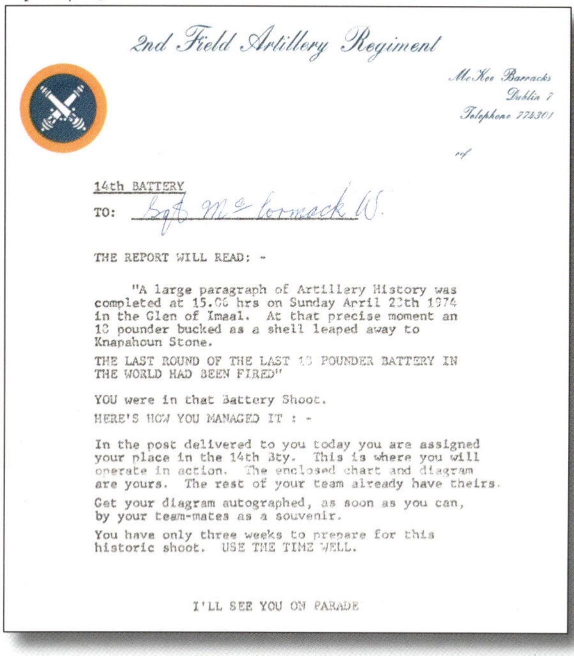

2nd Field Artillery Regiment

McKee Barracks
Dublin 7
Telephone 774301

14th BATTERY

TO: Sgt McCormack W.

THE REPORT WILL READ: -

"A large paragraph of Artillery History was completed at 15.06 hrs on Sunday April 28th 1974 in the Glen of Imaal. At that precise moment an 18 pounder bucked as a shell leaped away to Knapahoun Stone.

THE LAST ROUND OF THE LAST 18 POUNDER BATTERY IN THE WORLD HAD BEEN FIRED"

YOU were in that Battery Shoot.

HERE'S HOW YOU MANAGED IT : -

In the post delivered to you today you are assigned your place in the 14th Bty. This is where you will operate in action. The enclosed chart and diagram are yours. The rest of your team already have theirs.

Get your diagram autographed, as soon as you can, by your team-mates as a souvenir.

You have only three weeks to prepare for this historic shoot. USE THE TIME WELL.

I'LL SEE YOU ON PARADE

A letter from O/C, 14th Battery, FCA, describing the final shoot of the 18-pdr in Irish service; it notes that this was the last round of the last 18-pdr battery in the world to be fired. (Colin Stone)

The limber for an 18-pdr, upgraded with pneumatic tires, in The Square at Mullingar in September 2010. Two French 75s flank the flagpole in the rear. (Colin Stone)

Ordnance QF 25-pdr Mk III/1 and Mk IV

The 25-pdr Mk II Gun on Carriage 25-pdr Mark I was the standard British field gun of World War II. It was designed to replace the World War I vintage 18-pdr field gun and 4.5 inch howitzer, and is considered by some to be the best field artillery piece of the war. It entered into British service in 1940 and completed field service in the British Army in 1967, although it was used in a training role until the 1980s. In addition to being used by Commonwealth forces, it was used by a number of other Allied nations. In 1949 the Irish Army acquired forty-eight 25-pdrs; three variants (Mk II, Mk III, and Mk IV, all on the Mk I carriage) were in Irish service. At least some of the 25-pdrs delivered to Ireland had seen service during World War II, specifically, in North Africa. The Mk III had a modified receiver to prevent rounds slipping back when loading at a high angle, and the Mk III/1 added a muzzle brake. The Mk IV was a new-build version of the Mk III/1. The Mk I carriage had an integral firing platform that could be lowered, allowing the gun to be pulled on to it and quickly rotated in any direction. Each 25-pdr is accompanied by a limber that carries 32 rounds of ammunition, as well as tools and ancillary equipment. The ammunition for the 25-pdr is unfixed (separate loading), i.e., the shell itself is first loaded and rammed, and then the brass cartridge casing with the propellant charge is loaded and the breech closed for firing. The Irish Army fired the 25-pdr for the first time in March 1949 at the Seskin gun position in the Glen of Imaal, but the first regimental shoot of the 25-pdr in Ireland was on 21 June, 1949, also at the Glen of Imaal. The 21 June shoot consisted of 24 guns (three batteries) placed beside the Little Slaney River, and consisted of a number of different firepower demonstrations, including a rolling barrage, a regimental concentration, and an eight-gun battery firing in support of an infantry attack.

With a properly trained crew, the gun can be brought into action in two minutes. The six man gun crew consists of a No 1, in command of the gun crew; No. 2 operates the handbrake, rammer, and opens the cradle; No. 3 lays and fires the weapon, and fits the firing mechanism; No. 4, the loader, shows the cartridge case and charge to the No. 1 to confirm the charge; Nos. 5 and 6 prepare the ammunition (no. 6 also acts as driver for the towing vehicle). No. 2 through No. 6 all assist in lowering the gun platform. When moving out, No. 3 and No. 2 kick the A-frame out of its locked position. First- line ammunition load consists of 192 HE, 20 smoke, 14 HESH, and 10 illuminating rounds. Second line ammunition was 96 HE rounds. In 1998, thirty-six of the weapons delivered were still in frontline operational service (six with the PDF 2nd Field Artillery Regiment as a ceremonial battery, and the remainder with RDF field artillery regiments). The Artillery School

A somewhat "shop worn" 1 Battery, 62 RAR, 25-pdr during training in at the Curragh in 2009. The presence of female crew – on the left – is not unusual in the Irish Artillery Corps. (Colin Stone)

The last 25-pdr round being fired by 14th Battery, 7th FAR on 12 July 2005. The crew consists of Cpl S. Stein, Cpl C. Stone, RQ A. Quinn, Sgt M. Breen, and Cpl A. Doyle. (Colin Stone)

25-pdr ammunition, clearly illustrating the nature of the separate loading ammunition, consisting of the projectile itself, and a brass casing containing a separate propellant charge.(Bryan Ely)

A 25-pdr in a field position, well camouflaged with netting and branches. The crew appear ready to fire, although there is no evidence of either ready ammunition or expended shell casings. (MA, Artillery Box)

A 62 RAR 25-pdr firing during the 2008 camp. The stability provided by the lowered circular firing platform can be easily appreciated in this photograph. The platform also enables the entire gun to be traversed quickly with relative ease, as the tires ride on the circumference of the platform. The box trail allows sufficient space for recoil of the gun when the muzzle is at an elevated angle. (Colin Stone)

A 25-pdr at camp, Glen of Imaal, in 1995. Until definitive elimination from the active artillery inventory in 2009, the 25-pdr, along with the 120mm Brandt mortar before it was retired, was the mainstay of the Artillery Corps reserve units. (Colin Stone)

A battery of 2nd FAR 25-pdrs at McKee Barracks, Dublin. The guns are lettered A through F on the lower corner of each shield. The ACMAT VLRA gun tower for the second gun from the right is missing from the lineup. (RSM Mick Dillon via Colin Stone)

did not have its own 25-pdrs for training, but borrowed guns from 62 Reserve Artillery Regiment across the road from the Artillery School.

In the 1980s, close inspection using ultraviolet technology had revealed minute cracks in the breechblocks and breech rings on some of the 25-pdrs. The situation was remedied by replacing the breech rings and breechblocks by ones that were obtained from Canada. By late 2008, the useful life of the 25-pdrs was nearing an end for a number of reasons. Ammunition stocks had begun to run low, with little possibility of acquiring more; the latest stocks had been acquired from Pakistan Ordnance Factories. The age of the guns themselves had begun to tell as well, as the wheel bearings on the carriages had reached the stage that they were beyond replacement, and fractures were found in the towing eyes, with the result that towing them was no longer practicable, forcing the guns to be transported on low loader trucks, an equally unsatisfactory arrangement. Once they had arrived at Coolmoney Camp, they were allowed to be towed to their firing positions in the Glen of Imaal. The tires on the 25-pdrs represented an additional problem area; the Dunlop tires on all of the guns were original World War II tires, and somewhere in the very late 1990s or early 2000, someone determined that the tires should be replaced. The replacement effort was ill-starred; a trial gun was sent to Clancy Barracks to have its tires removed, but the tire machine there could not get the tires off. The tires were then sent to the Cavalry workshop at Cathal Brugha Barracks where there was another machine that supposedly could deal with the 25-pdr rims, as the tires proved to be too thick. The tires and rims were then outsourced to a civilian tire shop which used a team of four men and a 6-foot tire iron, all to no avail. The firm contacted the Irish Defence Forces to obtain permission to cut the tires, which were found to be three inches thick and still in perfect condition. However, it seems that a decision had been made at some level to replace all the tires, regardless, and forklift tires were obtained from South Africa, but they proved to be unsatisfactory in service, as the sidewalls were not as thick as the original Dunlops, and when the tires dropped into a rut, they tended to fail. Until mid-2009 each of the RDF artillery batteries were equipped with four 25-pdrs, but during that timeframe, the guns began to be phased out of service. On 22 July 2009, 54 Field Artillery Regiment fired its last shoot from the Stranahealy gun position; 62 Reserve Artillery Regiment fired its final regimental shoot with 25-pdrs on Wednesday, 22 July 2009. The last firing of an Irish 25-pdr took place on Thursday, 30 July 2009 at the Glen of Imaal. The shoot consisted of six guns (one from each artillery regiment), and the final shot was fired at 1500 hours by the Artillery School, with the Director of Artillery, Col Ray Quinn acting as the gun commander. A ceremonial battery of six Mk IV 25-pdrs for ceremonial occasions, outfitted with full inventory of accessories and accoutrements, is maintained at McKee Barracks, Dublin, by 2nd

Towing a 25-pdr – the hard way. Although relatively easy to move by a full crew on a level, paved surface, it is problematic as to how difficult such a move might be under field conditions. (Department of Defence)

An 8x8 IVECO Astra M320.42 carrying a 25-pdr to the final 25-pdr shoot on 30 July 2009. The Astra is one of at least four delivered to the Irish Army in 1997. The 25-pdr is fully tarped to protect it from the elements, although its left tire seems to be in need of inflation. The Astra was used only exceptionally to carry the 25-pdr portee-style. (Colin Stone)

Field Artillery Regiment. The 25-pdr has the distinction of being the longest-serving field artillery gun type in the Irish Army; it served faithfully and well for a full sixty years in an active role (until 2009), longer than in any other army, and likely will serve for a number of years into the future in a ceremonial role.

A pair of 25-pdrs during the final 25-pdr shuut un 30 July 2009. The gun in the foreground is crewed by personnel from one of the PDF batteries, as evidenced by their black berets, while the crew of the second gun, wearing green berets, are from an RDF battery. It appears that there is at least one female, holding the brass propellant case, in the PDF crew, and possibly another female, holding the projectile, in the RDF crew. Women have been serving as gunners since the early 1990s. (Irish Defence Forces)

A 25-pdr during the final 25-pdr shoot on 30 July 2009. The shell casing can be seen ejecting from the breech. The 25-pdr served longer than any other artillery piece in the Irish Army, from 1949 to 2009. (Irish Defence Forces)

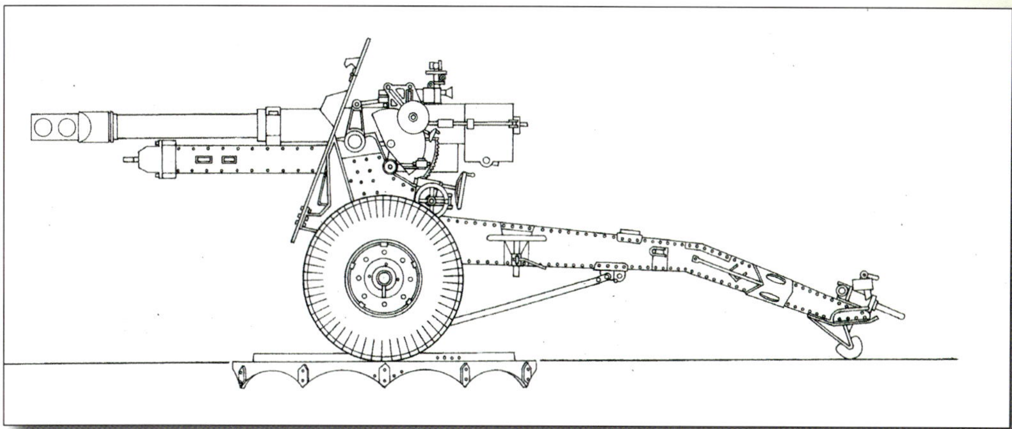

Designation	Ordnance QF 25-pr Mk III/1 and Mk IV
Manufacturer	Royal Ordnance Factories
Dates in Irish service	1949 – 2009 (12 retained as ceremonial guns after June 2009)
Number in Irish service	48
Caliber	3.45 inches (87.6mm)
Crew	6
Elevation	-5° to +45°
Traverse	360° on platform; 4° left or right on carriage
Weight	3968 lbs (1800 kg)
Length	181 inches (4602mm)
Width	84 inches (2134mm)
Height	65 inches (1650mm)
Carriage	Box trail
Barrel length	97.45 inches (2475mm)
Breech	Vertical sliding block
Recuperator	Hydro-pneumatic
Ammunition types	HE, HESH, smoke, illuminating, shot (blank)
Shell weight	25 lbs (11.4 kg) HE; 12.8 lbs (5.8 kg) smoke
Weight, complete round	30,9 lbs (14 kg) HE; 19.8 lbs (9 kg) smoke; 7.7 lbs (3.5 kg) blank
Muzzle velocity	Basic charge: 650 fps (198 m/s), Charge 2: 1007 fps (307 m/s), Charge 3: 1499 fps(457 m/s); Super charge: 1775 fps (541 m/s)
Rate of fire	6-8 rounds per minute
Range	Basic charge: 3609 yards (3300 m); Charge 2: 6562 yards (6000 m); Charge 3: 10936 yards (10,000 m); Charge Super: 13,123 yards (12,000 m)

The last round being fired by a 25-pdr in Ireland, 30 July 2009, at the Glen of Imaal. The gun commander was Col Ray Quinn, Director of Artillery, the rammer was Lt Col Edmond Fogarty, Commandant of the Artillery School, the layer was RSM John "Mingy" Morris, and the loader was Sgt Callery. (Lt Col Edmond Fogarty)

2nd FAR ceremonial 25-pdr battery firing a salute in Phoenix Park, Dublin. Following the retirement from service of the 25-pdr guns, long a hallmark of the Irish Artillery Corps, 2nd FAR has retained a battery of the guns for ceremonial purposes. (RSM Mick Dillon 2 FAR via Colin Stone)

3.7 inch Mountain Howitzer

The 3.7 inch mountain howitzer was first introduced in British service in 1917, and in addition to serving in World War I, also saw service during World War II. It was designed to be broken down into eight loads that could be packed by mule; it was also the first British artillery piece to use a split trail. A practiced gun crew could unload the gun from mules and assemble it in four minutes. Ammunition was separate loading, as described for the 25-pdr. Four of the pack howitzers were introduced into Irish service in 1933, and along with a limited number of 60mm and 81mm mortars then available formed the 1st Light Battery. There was no need to use the howitzer as a pack weapon in Irish service; it was used as a normal towed weapon. The 3.7 inch howitzer remained in the active inventory until the late 1970s, and was fired for the last time in 1977, during a course for newly commissioned artillery officers. The new officers conducted their terminal live firing practices firing 2,600 rounds of HE and WP during a four day firing exercise.

Volunteer Reserve personnel training on the 3.7" howitzer at Kildare barracks, possibly around 1935, under direction of Sgt McGough. This is only one of two photographs that have come to light of the gun is in its original configuration in Irish service, with the wood spoke wheels. (Mark McLoughlin)

A very unusual photograph of crew training on the 3.7" howitzer, specific date and location unknown, but likely at Kildare in the 1930s. It is highly unusual to see this gun fitted with the shield extensions that overlap the wheels. (Mark McLoughlin)

The Artillery Corps' 3.7 inch howitzer prior to its restoration, date unknown. Of the four guns of this type, one other is known to be in storage at McKee Barracks, and a third has been restored by the Cavan & Leitrim Railway, accounting for three-fourths the number of this type of weapon to have been on hand in the Irish inventory. (Sean Murphy)

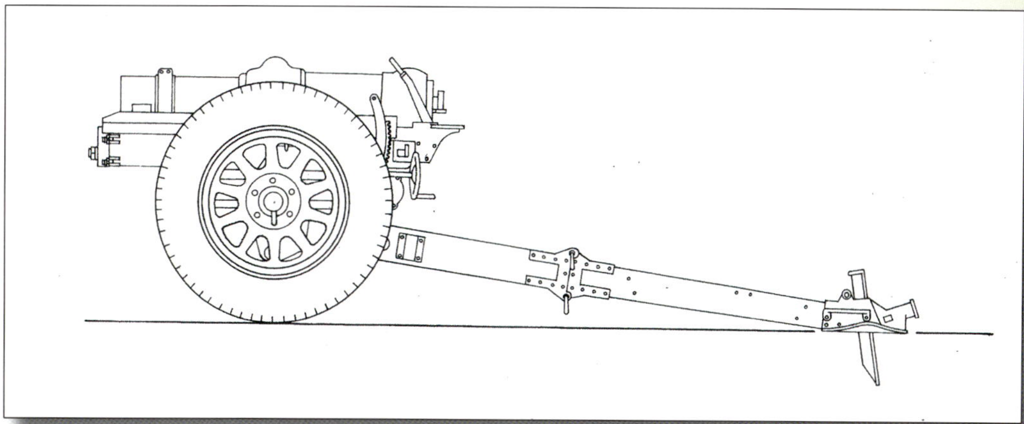

Designation	Ordnance QF 3.7 inch Mountain Howitzer
Manufacturer	Royal Ordnance Factory (?)
Dates in Irish service	1933-1977
Number in Irish service	4
Caliber	3.7 inches (94mm)
Crew	4
Elevation	-2.5° to + 42.5°
Traverse	20° left and right
Weight	1610 lbs (730 kg)
Length	133 inches (3383mm)
Width	8 inches (1473mm)
Height	6 inches (1422mm) to top of shield (approximate)
Carriage	Split trail
Barrel length	46.5 inches (1181mm)
Breech	Interrupted screw
Recuperator	Hydro-pneumatic
Ammunition types	HE, HEAT, WP, shrapnel, smoke, star shell
Shell weight	20 lbs (9.1 kg) (HE)
Muzzle velocity	973 fps (297 m/s)
Rate of fire	6-8 rounds per minute
Range	5899 yards (5394 meters)

A 3.7 inch mountain howitzer prepared to fire in the Glen of Imaal, 1976. This photograph offers a good view of the split trails, which seem especially long when compared to the howitzer itself. The gun commander has a Swedish Karl Gustav submachine-gun slung across his back. (62 Reserve Artillery Regiment, via Capt Ian Stewart)

A 3.7 inch mountain howitzer at a display on 7 May 2011. The gun has been restored by the Cavan and Leitrim Railway Museum, and is fitted with side shields not often seen while in service. This contrasts with photographs taken of the same howitzer in 2007, when the piece lacked any shield and was painted in different livery. (Colin Stone)

Breech of a 3.7 inch howitzer with breechblock open. The 3.7 inch howitzers in Irish service may have been the last guns of this type to have served in any army. (Sean Murphy)

The 3.7 inch mountain howitzer restored by the Cavan & Leitrim Railway Museum volunteers, shown on a Cavan & Leitrim Railway flat wagon in October 2007. Appropriate ammunition boxes complete the display. The absence of the gun shield allows a clear view of the extremely short length of the gun tube. This is one of three such howitzers remaining in Ireland, the others being in the care of the Artillery Corps. (Philip Bedford)

Gun, 105mm, Field, L118 and L119 (Light Gun)

In 1965 a British General Staff requirement for a new 105mm weapon system was approved. Prototypes of the new gun, designed by the Royal Armament Research and Development Establishment (RARDE) were tested in 1968. Following modifications made as a result of the tests, the gun was authorized for production in 1975, by Royal Ordnance Factory, Nottingham, and deliveries to the British Army began in 1976. In 1980, after considering adoption of the Italian OTO Melara 105mm pack howitzer, the Irish Army opted instead to purchase the British L118 Light Gun, largely because of its greater range capability. The first six-gun battery was delivered in 1980, and in May 1998, after a delay of 18 years, a second delivery brought the total to 24. The L118 has several interesting features. Its box trail is formed from tubular steel, and the gun tube itself can be rotated a full 180° in the travel position so that the muzzle faces rearward, shortening the overall length of the towed gun. The gun can also be towed in the normal configuration with the muzzle oriented in the firing position. The current Army configuration of three Regular brigades, each with an artillery regiment, accounts for eighteen deployed L118 guns (three batteries, one in each regiment, each with six guns). In addition to these eighteen L118s, there are six L119 Light Guns at the Artillery School dedicated to training courses, and familiarization and firing by the Reserve Defence Forces. The L119 has a shorter barrel than the L118, and is chambered to fire

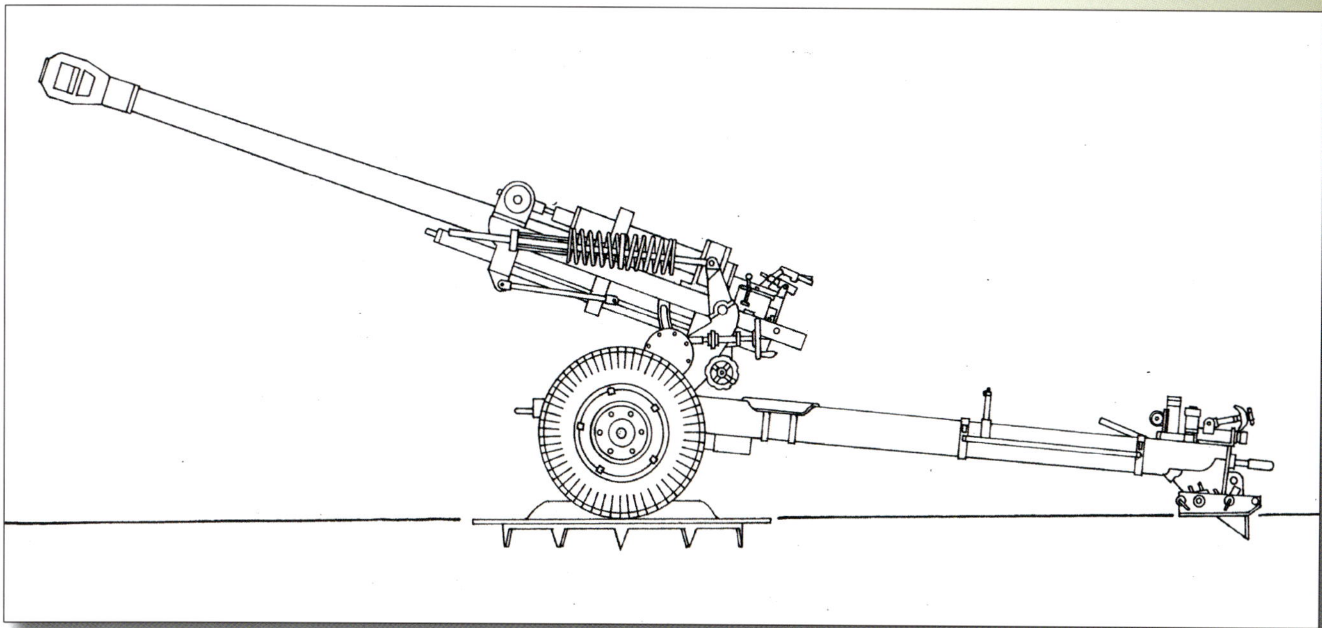

105 light guns firing, Glen of Imaal, 28 March 1985. These particular guns were from the first tranche of L118s delivered in 1980, and were assigned to the 2nd Field Artillery Regiment. (Denis McCarthy)

An L118 light gun in July 2007. The tires are resting on a firing platform similar to that of the 25-pdr, allowing the entire gun to be traversed easily. The truck behind the gun is a Leyland DAF, often used as an artillery prime mover. (Irish Defence Forces)

A 2007 photo of an L118 providing an excellent view of the overall length of the gun and its low silhouette. The gun is resting on the classic British-style circular firing platform that provides stability. (Colin Stone)

Firing the L118. A spent casing is just clearing the breech, and another round is ready to hand off to the loader. A well trained, proficient crew can fire six to eight rounds per minute. (Department of Defence)

A 105mm light gun firing in the Glen of Imaal on 30 July 2009. The L118 equips one battery in each of the Irish Army's three Regular artillery regiments as well as a school battery. (Irish Defence Forces)

L118 105mm light gun with gun reversed for towing, 24 June 2007. The size of the tubular split trail with respect to the gun tube itself can be appreciated in this photograph. (Paul McMenamin)

the more economical US M1 type ammunition rather than the separate loading ammunition used in the UK 105mm howitzer. None of the L 118s are on issue to reserve artillery units.

Designation	Gun, 105mm, Field, L118 (more commonly referred to as the Light Gun)
Manufacturer	Royal Ordnance Factory (ROF), Nottingham
Dates in Irish service	1980 – present
Number in Irish service	24 (18 L118, 6 L119)
Caliber	4.13 inches (105mm)
Crew	6
Elevation	-5.5° to +70°
Traverse	360° on platform; 11° on carriage
Weight	4096 lbs (1858 kg)
Length	346.5 inches (8800mm)
Width	70 inches (1780mm)
Height	84 inches (2130mm)
Carriage	Box trail
Barrel length	148 inches (3750mm)
Breech	Vertical sliding block
Recuperator	Pneumo-hydraulic (recuperator, recuperator reservoir, buffer cylinder,right and left compensating cylinders)
Ammunition types	HE, HESH, smoke, illuminating, target marking
Shell weight	33.3 lbs (15.1 kg) HE; 23 lbs (10.49 kg) HESH
Muzzle velocity	2326 fps (709 m/s)
Rate of fire	6-8 rounds per minute
Range	18,810 yards (17,200 meters) (HE); 875 yards (800 meters) in anti-tank mode

A 105mm light gun under tow by a Leyland DAF 4x4 TCV at a display of Irish Army equipment, June 2007. Two Mowag Piranhas and a Scorpion CVR(T) are in the background. (Paul McMenamin)

A 105mm L118 howitzer at a military display at Salthill, Galway. While onlookers brave the elements, the gunners seem quite comfortable in the back of their Leyland GTV. (Roy Kinsella)

One of the six L119 light guns undergoing maintenance at Curragh Camp, 27 May 2009. The L119s fire a less expensive type of 105mm round than does the L118, and are assigned to the Artillery School. The tires on this gun appear brand new. (Ralph Riccio)

Ordnance QF 4.5 inch Howitzer

The 4.5 inch howitzer was the standard British field howitzer in World War I. The Mark I version was introduced in 1909, and in 1917, the Mark II version was introduced. The howitzer originally had wooden spoke wheels, as was standard for the era, but in the 1920s the carriages were upgraded with steel wheels and pneumatic tires to allow towing by motorized vehicles. In 1925

A 4.5 inch howitzer in original configuration, with wood spoke artillery wheels. The crew are wearing the 1926 helmet that was made by Vickers on machinery received as war reparations that had previously been used to produce the German M1916 "coal scuttle" helmet. The combination of the Vickers helmet and the wooden wheels dates the photograph to the time frame between the late 1920s and 1940, by which time the Irish Army had upgraded all of its guns with pneumatic tires suitable for mechanical towing. (Colin Stone)

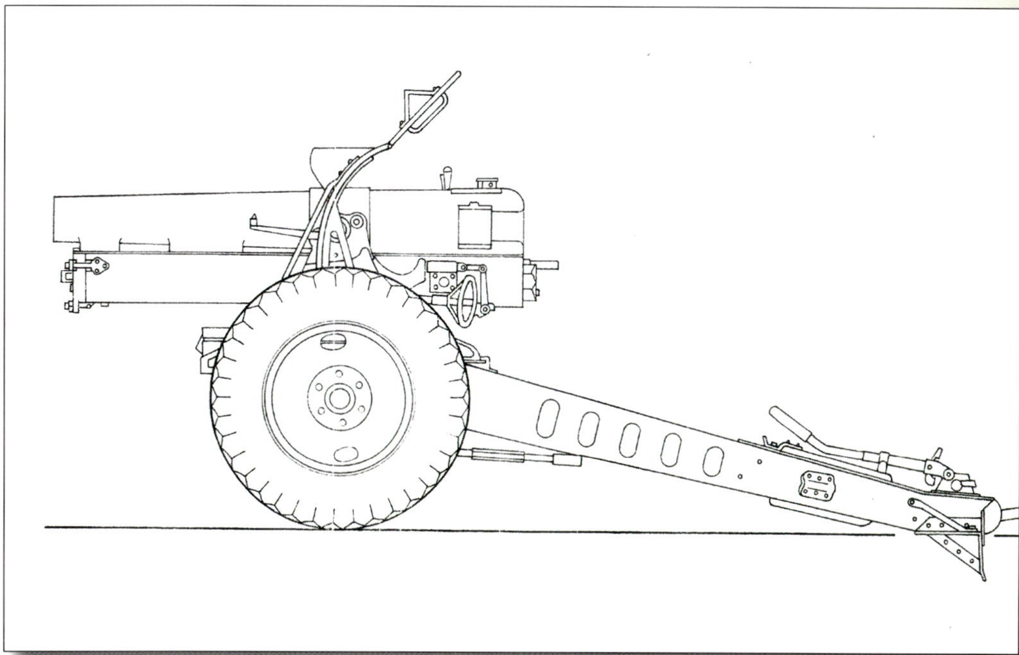

Ireland received four of the 4.5 inch howitzers, and by 1949 a total of 38 had been acquired. The 4.5 inch howitzers in Irish service all appear to have been converted using a Martin Parry carriage. They equipped some FCA artillery units until as late as 1976.

A 4.5 inch howitzer ready to fire. The upper part of the shield is lowered, allowing the crew an unrestricted view to the front of the gun. The 4.5 inch howitzer and the 18-pdr gun were the mainstays of the Irish Artillery Corps during The Emergency. (Sean Murphy)

A 4.5 inch howitzer participating in Exercise Keadeen 1966. The crew consists of, from left to right: Gunners J. Geohagan, E. Foran (partially obscured), M. Dowd, and J. Boyce. Missing from the picture is the gun commander, Sgt. P. Dooley. Although the howitzer is in a carefully prepared revetted position, there is no visible evidence of ammunition ready at hand (MA, Artillery Box)

A lineup of 4.5 inch howitzers at Coolmoney Camp, 13 July 1975. Two guns are uncovered, while the leftmost gun is covered by a protective tarp. The middle gun has a different axle hub profile. (Denis McCarthy)

A pair of 4.5 inch howitzers at Coolmoney Camp, 13 July 1975. The two guns have different style axle hubs; the closer gun has the more commonly encountered style of hub, while the gun on the right has a hub with a ring that allows ropes or a chain to be attached for pulling. (Denis McCarthy)

This 4.5 inch howitzer was restored by the Artillery Corps; as shown here, the restoration is nearly complete. The gun shield has yet to be fitted, although the shield support brackets are in place. (Sean Murphy)

The fully restored 4.5 inch howitzer, complete with the raised upper shield. This is one of several 4.5 inch howitzers to have been preserved in Ireland. (Sean Murphy)

Designation	Ordnance QF 4.5 inch Howitzer
Manufacturer	Coventry Ordnance Works
Dates in Irish service	1925-1976
Number in Irish service	38
Caliber	4.5 inches (114.3mm)
Crew	6
Elevation	-5° to +45°
Traverse	3° right and left
Weight	3291 lbs (1493 kg)
Length	147 inches (3733mm) travel order, 159 inches (4039mm) firing position
Width	78.25 inches (1988mm)
Height	58 inches (1473mm)
Carriage	Box trail
Barrel length	70 inches (1778mm)
Breech	Horizontal sliding block
Recuperator	Hydro-spring
Ammunition types	HE, shrapnel, smoke, star, incendiary, gas
Shell weight	35 lbs (15.9 kg) (HE)
Muzzle velocity	1010 fps (308 m/s)
Rate of fire	4 rounds per minute
Range	7300 yards (6675 meters)

Ordnance BL 60 pdr Mk 1

The Ordnance BL 60-pdr was designed in 1904, and remained in service through WW II. It is variously referred to as a medium gun and a heavy gun. As the BL designation implies, it was a breech loading gun firing separate ammunition (i.e., the projectile and the propellant charge were two separate items, the propellant being encased in cloth bags rather than in a brass shell casing). In terms of caliber, these were the largest field guns ever in Irish service, and were also the heaviest in terms of weight (coast artillery guns excepted). Due to their size and weight, the first time the

BL 60-pdr gun drill at Collins Barracks, Dublin, in 1943. The 60-pdr was a massive gun, and in terms of bore diameter, was the heaviest mobile gun ever fielded by the Irish Army. It was not upgraded with pneumatic tires, retaining the impressively large World War I style wood spoke artillery wheels. (Brian Lowen)

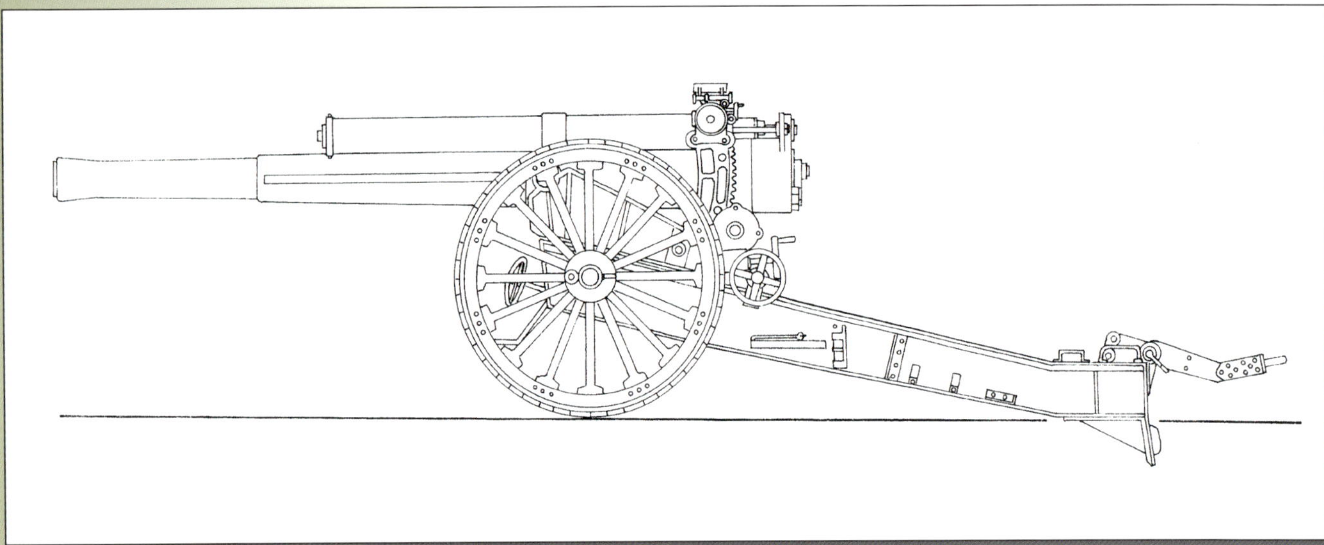

Members of the 41st/42nd Btys (later 14th/19th Btys) ready to fire a 60-pdr in the Glen of Imaal in 1942. (Shay Stone via Colin Stone)

Laying a 60-pdr for firing at the Glen of Imaal, with other crew members in their assigned positions. Because of their size, weight, spoke wheels and solid tires, towing the guns over any distance would have been problematic. (MA, via Lt Col Edmond Fogarty)

The layer for a 60-pdr signaling "trails left" to instruct the crew to move the gun carriage to the right. Collins Barracks, Dublin, probably in 1943. (Shay Stone via Colin Stone)

60-pdrs were sent to the Glen of Imaal for an annual camp, it took two days to get the guns from Collins Barracks in Dublin to the Glen, loaded aboard an ESB (Electric Supply Board) low loader. Although there are some references to the 60-pdr being used as a coast defense gun in Irish service, they were issued to the 41st and 42nd Field Batteries, LDF. Four were still reflected in the artillery inventory as of 1949, but were not assigned as battery equipment at the time. It is not clear when the 60-pdrs were boarded; none have survived.

Designation	Ordnance BL 60 pounder gun (Mk II)
Manufacturer	Elswick Ordnance
Dates in Irish service	1942-1949?
Number in Irish service	6
Caliber	5 inches (127mm)
Crew	10
Elevation	-4° to + 35°
Traverse	4° right and left
Weight	12,048 lbs (5465 kg)
Length	192.25 inches (4883mm)
Width	NA
Height	NA
Carriage	Box trail
Barrel length	185 inches (4699mm)
Breech	Asbury single-motion
Recuperator	Hydro-pneumatic variable
Ammunition types	HE, shrapnel, smoke, gas
Shell weight	56 lbs (25.4 kg)
Muzzle velocity	2130 fps (649 m/s)
Rate of fire	2 rounds per minute
Range	15,500 yards (14.173 meters)

Anti-tank Artillery

Ordnance QF 2 pounder

The Ordnance QF 2-pdr was originally designed as a tank gun, but in response to a 1934 Staff requirement it was accepted as the basis for an anti-tank gun. A limited production run ensued in 1936, and the gun entered service beginning in 1938. The gun was adopted as the Ordnance QF 2-pounder Mk IX on Carriage Mark II. Large numbers of the gun were left behind at Dunkirk; it was used extensively in the Western Desert campaign, but it proved to be relatively ineffective because of its high silhouette and lack of range compared to opposing tank guns. The carriage was an unusual design consisting of three legs, one of which was used for towing in the travel mode while the other two were folded; the carriage has two rubber-tired wheels for towing. When the gun was deployed for combat, the legs were placed on the ground and

The 2-pdr gate guardian at Cathal Brugha Barracks, Dublin, on 5 May 2011. This is one of only seven 2-pdrs to have been in Irish service early during The Emergency period. The outriggers for the cruciform platform are shown extended in the front of the gun, but would be folded back while in the travel position. (Colin Stone)

A ¾ rear view of the 2-pdr at Cathal Brugha Barracks. The gun has been raised off the ground with blocks placed under the axle in order to relieve stress on the tires. (Colin Stone)

A rather freshly painted 2-pdr on its cruciform platform, with wheels removed, shown at the Curragh in November 2011. This gun was previously located at Columb Barracks, Mullingar. (Colin Stone)

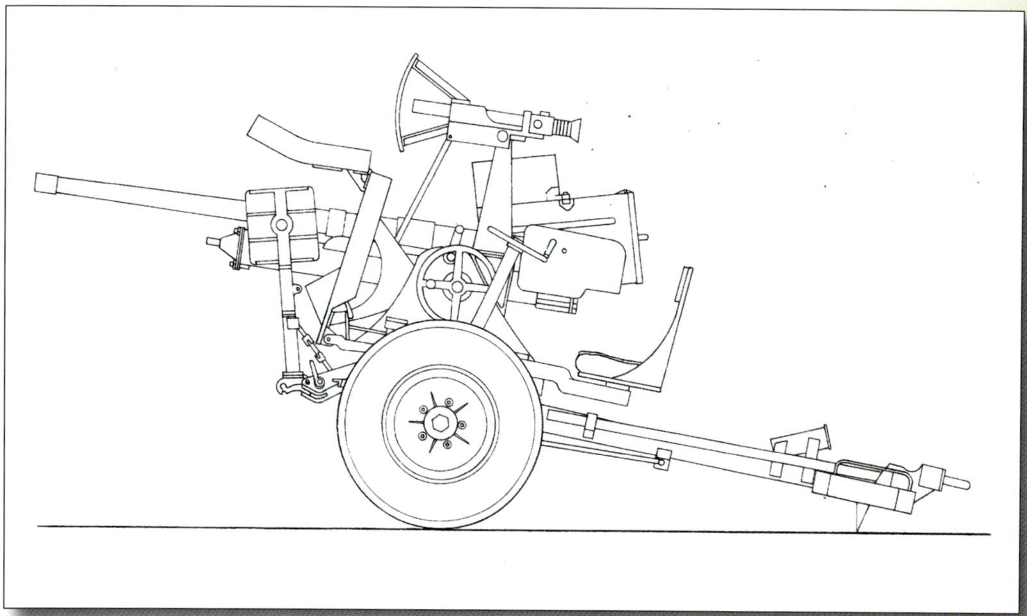

the wheels folded up. This arrangement provided good stability for the gun and enabled it to be swung through a complete 360° arc for firing. In October 1939 Ireland received one Mk II 2-pdr, and in November 1942 received another six of the 2-pdr AT guns. The 2-pdrs reportedly were issued to the infantry regiments following The Emergency, and remained in the inventory until about 1959. Four examples of Irish 2-pdrs seem to have survived, one at Cathal Brugha Barracks, Dublin, one at the Curragh, and a third at Collins Barracks, Cork; a fourth was undergoing refurbishment at Curragh Camp in December 2011.

Designation	Ordnance QF 2 pounder
Manufacturer	Vickers-Armstrong
Dates in Irish service	1939- 1959 (?)
Number in Irish service	7
Caliber	1.57 inches (40mm)
Crew	3
Elevation	-13° – +15° (-5° to + 25° when on wheels)
Traverse	360° (14° left, 10° right when on wheels)
Weight	1847.5 lbs (838 kg)
Length	135.5 inches (3442mm)
Width	55.15 inches (1401mm)
Height	NA
Carriage	Three folding legs; extended to form a platform for firing
Barrel length	82 inches (2083mm)
Breech	Vertical sliding block
Recuperator	Hydrospring
Ammunition types	AP
Shell weight	2 lbs (0.91 kg)
Muzzle velocity	2600 fps (792 m/s) (AP-T and HE-T)
Rate of fire	12 rounds per minute
Range	1000 yards (914 meters) effective; 8000 yards (7315 meters) maximum

The 2-pdr maintained at Collins Barracks, Cork, photographed in November 2011. Note the lowered top portion of the shield. Visible in the background, somewhat protected from the elements by an overhead covering, is the museum's restored Leyland Retriever machinery lorry. (Colin Stone)

A very poor quality photograph of officers training on the 2-pdr gun at what appears to be an indoor range in 1945. The wheels have been removed so that the gun can rest on its integral stabilized platform. (Irish Defence Forces via Colin Stone)

Ordnance QF 6 pounder 7 cwt Mk 2

The Ordnance QF 6 pounder was developed at Woolwich Arsenal between 1938 and 1940, as it was apparent that the 2-pdr anti-tank gun was inadequate to cope with the threat faced by British forces. The carriage for the gun was not ready until 1941, however, and due to several factors the first 6-pdrs were not issued until May 1942. The 6-pdr was the primary British anti-tank gun from that point to the end of the war, and a copy was also used in U.S. service as the 57mm Gun M1. In about 1949 Ireland acquired six Mark II 6-pdr anti-tank guns that served until about 1959.

A 6-pdr anti-tank gun at Cathal Brugha Barracks, 31 August 1998, at a display marking the 50th anniversary of the "End of The Emergency." The white muzzle brake and the red highlights on handles and tow hooks were not normal livery for guns in regular service, which were painted a uniform green or olive drab. (Denis McCarthy)

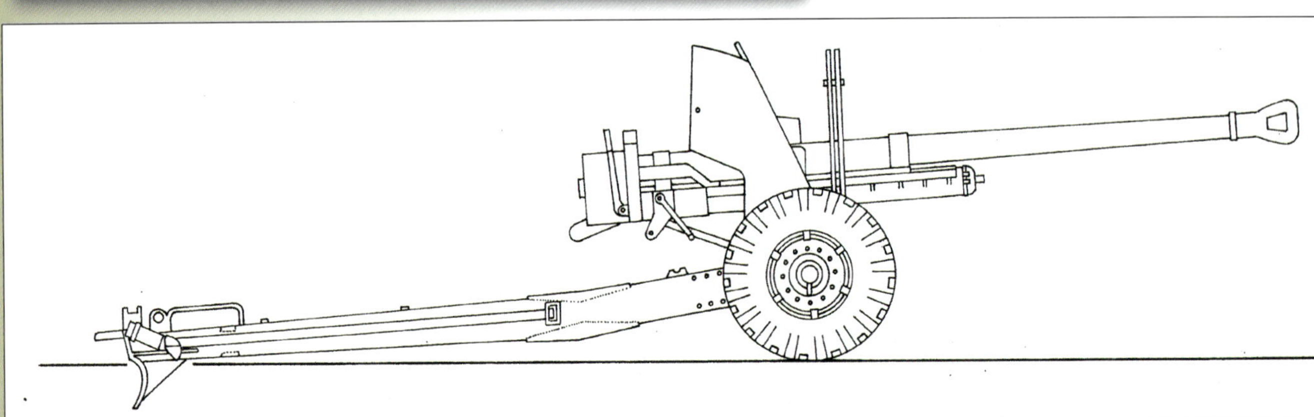

CDSW ZC 5831 towing a 6-pdr anti-tank gun on a country lane. The canvas tilt is in place to protect against the weather, and some attempt has been made at camouflaging the vehicle by placing branches on the forward portion of the vehicle. (MA Artillery Box)

The business end of a 6-pdr, photographed at Curragh Camp, 27 May 2009. The British 6-pdr was adopted by the U.S. Army in 1942 as the 57mm Gun M1, differing from the British original in some details such as carriage and shield configuration. (Ralph Riccio)

Right profile of the 6-pdr at Curragh Camp in May 2009. Although large numbers of the 6-pdr were issued to British and Commonwealth units during the war, Ireland had to wait until the war was over to acquire six of the 6-pdrs as surplus equipment. (Ralph Riccio)

The very colourful Cathal Brugha Mark II 6-pdr seen from the rear. Note the mismatched tires. (Colin Stone)

The 6-pdr maintained at Cathal Brugha Barracks, 5 May 2011. Red paint has been liberally applied to the muzzle brake and to many other detail surfaces of the gun to make it appear more appealing, at least to some eyes. (Colin Stone)

Designation	Ordnance QF 6 pounder, Mk IV
Manufacturer	Woolwich Arsenal (?)
Dates in Irish service	1949 – 1959 (?
Number in Irish service	6 (both Mk II and Mk IV were in service, but breakout is not available)
Caliber	2.24 inches (57mm)
Crew	5
Elevation	-5° to +15°
Traverse	90° (45° left and 45° right)
Weight	2520 lbs (1143 kg)
Length	185.25 inches (4705mm)
Width	71.5 inches (1816mm)
Height	50.5 inches (1283mm) (to top of gun shield)
Carriage	Split trail
Barrel length	112.2 inches (2850mm) Mk IV; 96.3 inches (2447mm) Mk II
Breech	Vertical sliding block
Recuperator	Spring recuperator with hydraulic buffer
Ammunition types	AP, APC, APCBC, APCR, APDS, HE
Shell weight	6.25 lbs (2.8 kg) AP
Muzzle velocity	2639 fps (820 m/s)
Rate of fire	12 – 15 rounds per minute
Range	500 yards (5030 meters); effective range against tanks 800 yards (732 meters)

Ordnance QF 17 pounder Mk I

Even before the 6-pdr entered service, the British Army realised that it would not be able to cope with the developing German threat, and accordingly, in 1940 initiated a design effort to develop a replacement. The first 17-pdrs, provisionally mounted on 25-pdr carriages, initially saw action in February 1943. Full production began in 1943. The gun could penetrate 130mm of armour, sloped at a 30 degree angle, at 1000 yards. The gun was much heavier and bulkier than its predecessor (the 6-pdr) and could not be manhandled into position by its crew, and because of the need for a suitable prime mover, was issued only to anti-tank and armoured units. Following WW II a glut of surplus equipment on the world market enabled Ireland to acquire a significant amount of newer, more capable artillery; included in this were twelve British 17-pdr anti-tank guns, purchased somewhere around 1949 or 1950. The guns were assigned to anti-tank batteries of the three field artillery brigades, but later appear to have been transferred to equip the anti-tank troops of the Reserve batteries of the anti-aircraft regiment. The 17-pdr was a much more powerful gun than

A ¾ right front view of a 17-pdr in Irish service, date unknown. The 17-pdr was a potent anti-tank weapon but it was not until after the war was over that Ireland acquired twelve of these guns at bargain prices. (Sean Murphy)

The 17-pdr with its split-trail carriage opened as it would be for firing. The date of this photograph is uncertain. (Sean Murphy)

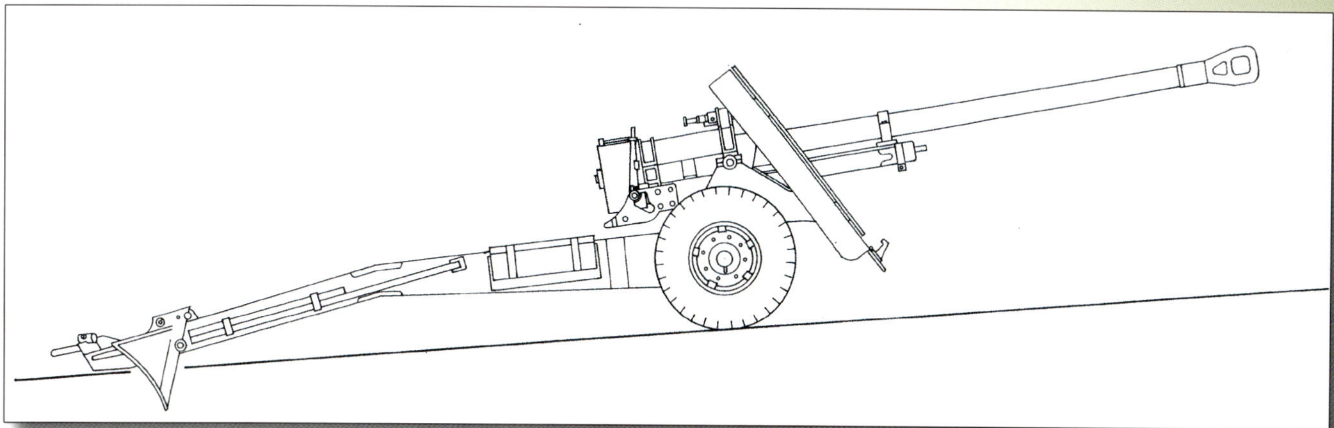

A 17-pdr anti-tank gun at Cathal Brugha Barracks, 31 August 1998, at a display marking the 50th anniversary of the "End of The Emergency." The camouflage pattern is not known to have been standard livery while the 17-pdr was in active service in the Irish Army.(Denis McCarthy)

A 17-pdr at Columb Barracks, Mullingar, Artillery Day 1998. The gun has been lifted from the ground and is resting on jacks; despite the care which has been taken to restore the gun, water has settled in the wheel rim, causing surface rust. (Colin Stone)

One of the two 17-pdrs maintained at Cathal Brugha Barracks, Dublin, in a 5 May 2011 photograph. The gun presents a rather massive appearance, and indeed, was a fearsome anti-tank weapon for its time. (Colin Stone)

Rear view of 17-pdr showing breech mechanism, elevation and traverse handwheels, sighting apparatus, and trails opened. The double-skinned shield provided protection against small arms fire and shell fragments. (Colin Stone)

earlier anti-tank weapons in the Irish inventory, and a somewhat apocryphal story says that when the 17-pdr was first fired in Ireland, the first four rounds crossed over the next glen and landed in the Irish Sea! They likely were phased out of service around 1962. At least four have survived: two at Cathal Brugha Barracks, Dublin, one at Columb Barracks, Mullingar, and the other at a private museum.

Designation	Ordnance QF 17 pounder
Manufacturer	Royal Ordnance Factories (?)
Dates in Irish service	1949 – 1962 (?)
Number in Irish service	12
Caliber	3 inches (76.2mm)
Crew	5
Elevation	-6° to +16.5°
Traverse	60° (30° right and left)
Weight	4625 lbs (2098 kg)
Length	175 inches (4443mm)
Width	NA
Height	NA
Carriage	Split trail, with gun shield
Barrel length	140 inches (3556mm)
Breech	Vertical sliding block
Recuperator	NA
Ammunition types	AP, APC, APDS, HE
Shell weight	17 lbs (7.65 kg)
Muzzle velocity	3950 fps (1204 m/s) (APDS)
Rate of fire	10 – 12 rounds per minute
Range	10,000 yards (9144 meters)

The second of the two 17-pdr guns at Cathal Brugha Barracks, taken in May 2011. The widely spread split trails provided stability needed for the gun, whose recoil forces were considerable. The 17-pdr in Irish service was quite a radical departure in capability and performance over the relatively sedate 6-pdr and 2-pdr anti-tank guns previously in the inventory. (Colin Stone)

Mortars

Mortars represent the most numerous type of artillery in the Irish Army, accounting for over half of the aggregate number of all artillery types ever issued in the Irish Army. Mortars are an attractive option for Irish circumstances, as they are inexpensive, easy to operate, highly mobile, require little maintenance, and, especially in the case of the larger 120mm mortars, provide a respectable amount of firepower. In common with many other armies, mortars in the Irish Army are issued to infantry as well as artillery units. The 60mm, 3 inch, and 81mm mortars have all been organic to infantry units; prior to The Emergency the 81mm mortars were assigned to artillery batteries, but in 1940 they were transferred to the infantry. Since then, due to particular circumstances, 81mm mortars have at times been relinquished to the artillery. As described previously, mortars have represented the sole use of Irish artillery deployed overseas in support of peacekeeping operations.

Infantry use of mortars

The mortar is sometimes referred to in the U.S. Army as "the company commander's artillery." Although strictly speaking any mortar is an artillery weapon, the size, weight, portability and flexibility of mortars up to about 81mm historically have enabled them to be organic to infantry rather than artillery units. This axiom holds true in the Irish Army which has included mortars ranging from 60mm to 81mm in the infantry inventory. During The Emergency, the scale of issue was four medium mortars (3 inch Mk II or 81mm Brandt) per infantry battalion, although with a total of 98 medium mortars (fifty 3 inch and forty-eight 81mm) there may likely have been a slight shortfall in at least a few battalions. Currently (2012), infantry battalions are allocated four 81mm mortars and six 60mm mortars.

Mortier de 60mm Modèle 1935 (Brandt)

The 60mm Modèle 1935 was a product of the Edgar Brandt design bureau and entered French service in 1937 (it is sometimes referred to as the Model 1937). The U.S. copy of the 60mm Brandt was the M1 mortar, which later developed into the M2 mortar. Although unconfirmed, it is believed that this is the company-level mortar used by the Irish Army during The Emergency and probably until about 1981 when it was replaced by the newer MO-60-63.

Designation	Mortier de 60mm Modèle 1935
Manufacturer	Hotchkiss-Brandt, France
Dates in Irish service	1940-1981 (?)
Number in Irish service	Unknown
Caliber	60mm
Crew	5
Weight (complete)	43.4 lbs (19.7 kg)
Weight, barrel	12.8 lbs (5.8 kg)**
Weight, bipod	16.4 lbs (7.4 kg)**
Weight, base plate	12.8 lbs (5.8 kg)**
Barrel length	28.5 inches (724mm)
Ammunition types	HE, fragmentation
Shell weight	2.93 lbs (1.33 kg)
Muzzle velocity	518 fps (518 m/s)
Rate of fire	20-25 rounds per minute normal
Range	1860 yards (1700 meters)

** Weights shown are for U.S.M2 mortar.

Hotchkiss-Brandt MO-60-63 60mm mortar

The Hotchkiss-Brandt MO-60-63 60mm mortar entered Irish service, probably in 1981 along with the larger 81mm Brandt LMLB mortar, and was issued at company level. The MO-60-63 was

The 60mm Hotchkiss-Brandt mortar in its wooden carrying case. Weighing in at less than 40 pounds, along with its ammunition it can be easily carried by a small team of infantrymen along with front-line troops. (Darren Waters, via Roy Kinsella)

The Hotchkiss-Brandt 60mm mortar used by the Irish Army from about 1981 to 1997. The 60mm mortar was used exclusively by Irish infantry units and was not issued to the Artillery Corps. Its light weight makes it extremely suitable for accompanying infantry units and provides instantly available organic fire support for the infantry commander. (Darren Waters, via Roy Kinsella)

withdrawn from service in 1997 following an accident at Kilworth in which a female soldier lost her hand due to a misfire and barrel failure of the mortar as she was loading it during a training course.

Designation	60mm MO-60-63
Manufacturer	Hotchkiss-Brandt, France
Dates in Irish service	1981 (?) – 1997
Number in Irish service	Unknown
Caliber	60mm
Crew	4
Weight (complete)	39.5 lbs (17.9 kg)
Weight, barrel	NA
Weight, bipod	NA
Weight, base plate	NA
Barrel length	28.5 inches (724mm)
Ammunition types	HE, smoke, illumination
Shell weight	NA
Muzzle velocity	NA
Rate of fire	NA
Range	2187 yards (2000 meters)

The 60mm Denel Vektor M1 light mortar that entered Irish service in 2003. The Denel is issued to both Regular and Reserve units and can be fired both with and without the use of a base plate. (Irish Defence Forces)

Denel Vektor M1 60mm mortar

The Denel Vektor is a modern light mortar developed in South Africa. It entered Irish service in 2003, replacing the earlier 60mm Brandt mortar; the Army Reserve were equipped with it in 2005. The mortar consists of three main assemblies (the barrel, bipod, and base plate), but because of its light weight can be carried in a single unit. Its light weight enables it to be moved quickly and easily, providing readily available support as needed. The sight is similar to the C2A1 sight used on the 81mm mortars in Irish service, and is compatible with the Morfire fire control system used with the 81mm mortar. By removing the bipod and base plate, and attaching a special small base plate, and different sighting unit and a handgrip the mortar can be used in the commando role.

A 60mm Vektor mortar team with their weapon. The individual weapons slung over their shoulders are the standard issue 5.56mm Steyr AUG 1 Austrian assault rifles, introduced into Irish service in 1988. (Irish Defence Forces)

Preparing to fire the Vektor mortar. The light weight of both the mortar and its ammunition make it an ideal support weapon for light infantry units. Various models of 60mm mortars have been issued as organic weapons to Irish infantry units since about 1940. (Irish Defence Forces)

A 60mm Vektor at the moment of firing. The crew is able to fire the mortar from a seated position, as shown, rather than kneeling or standing as is obligatory with larger caliber mortars. (Irish Defence Forces)

Designation	Denel Vektor M1
Manufacturer	Denel Vektor, South Africa
Dates in Irish service	2003 – present
Number in Irish service	NA
Caliber	60mm
Crew	4
Weight (complete)	39.25 lbs (17.8 kg)
Weight, barrel	8.8 lbs (4.0 kg)
Weight, bipod	12.3 lbs (5.6 kg)
Weight, base plate	14.8 lbs (6.7 kg)
Weight, sight	3.3 lbs (1.5 kg)
Barrel length	25.6 inches (650mm)
Ammunition types	HE, smoke, illumination
Shell weight	NA
Muzzle velocity	97-597 fps (60-182 m/s)
Rate of fire	20 rounds per minute maximum
Range	2442 yards (2233 meters)

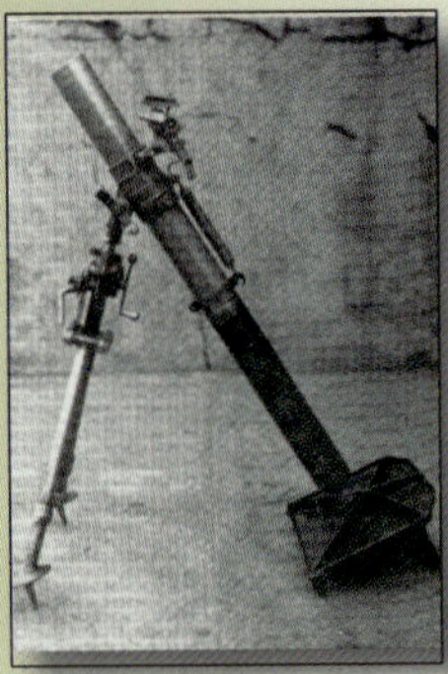

A British 3-inch mortar, copied from an Irish manual. The recoil spring about a third of the way down the barrel is a key recognition feature for this mortar. Fifty of these weapons were supplied to Ireland by Britain during The Emergency. (Manual of the Mortar)

Ordnance ML Mortar, 3-inch Mk II

The Ordnance ML 3 inch mortar Mk II superseded the Stokes 3 inch mortar in the British Army in the 1920s, and served until the 1960s as the standard infantry support weapon. It was a conventional muzzle loaded and drop-fired mortar. Britain supplied fifty of the ML mortars to Ireland in 1943/1944.

Designation	Ordnance ML 3 inch mortar Mk II
Manufacturer	?
Dates in Irish service	1943 – 1981
Number in Irish service	50 (?)
Caliber	3 inches (76.2mm)
Crew	3
Weight (complete)	126 lbs (57 kg)
Weight, barrel	44 lbs (20 kg)
Weight, bipod	45 lbs (20.4 kg)
Weight, base plate	37 lbs (16.8 kg)
Barrel length	46.85 inches (1190mm)
Angle of fire	+45° to +80°
Ammunition types	HE, smoke, practice, drill
Shell weight	10 lbs (4.54 kg)
Muzzle velocity	650 fps (198 m/s)
Rate of fire	NA
Range (maximum)	850 yards (777 meters) charge I; 1600 yards (1463 meters) charge II

Brandt 81mm mortar (M44)

In the 1930s the French firm of Edgar Brandt produced a widely accepted series of mortars; the Brandt name has become synonymous with mortars ever since. In 1936 the Irish Army obtained 31 Brandt 81mm mortars, presumably directly from the French manufacturer. According to Irish Ordnance Corps records, by 1942 there were 48 Brandts in the inventory. In the 1950s

An 81m M44 mortar detachment, likely training during The Emergency. As a mortar bomb is being readied to load into the muzzle, two other rounds are at hand to continue firing. (Military Archives via Denis McCarthy)

A five-man mortar detachment preparing to fire a Brandt M44 mortar, probably during The Emergency. The 81mm mortars initially were issued to Volunteer Force light artillery batteries but were transferred to the infantry in 1940. (Military Archives via Denis McCarthy)

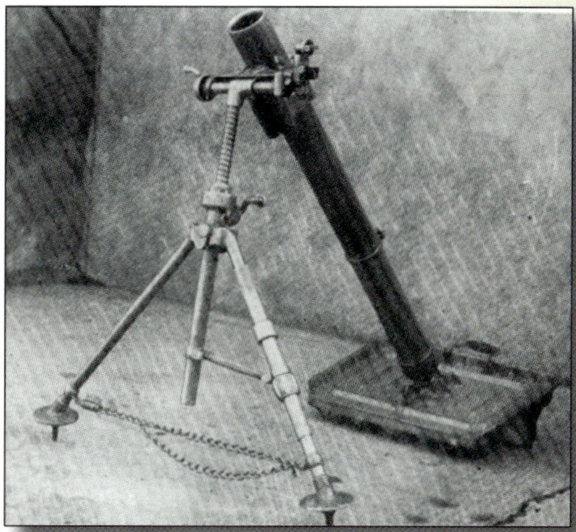

An 81mm Brandt M44 mortar. During The Emergency, 48 of these mortars were obtained by Ireland and were issued to infantry units. They were the heaviest mortars in the Irish inventory until the advent of the 120mm Brandt AM 50 mortar that entered Irish service in 1953. (Manual of the Mortar)

An illumination round being readied for firing in an 81mm Brandt M44 mortar with the Irish UNIFIL contingent in Lebanon in 1980. The 81mm mortar supplemented the larger Brandt 120mm mortar with the Irish contingent in Lebanon. (Martin Forde via Roy Kinsella)

an unspecified number of additional Brandt mortars were acquired. These 81mm Brandts in Irish service were referred to as the M 44. From 1936 until June 1940 these mortars were issued to Volunteer Force Light Batteries, but in June 1940 the light batteries were designated as mortar batteries, and their personnel and equipment were transferred to the infantry. There are some indications that these mortars were issued on a scale of two per battalion. In 1981 these mortars were replaced by the 81mm Hotchkiss Brandt MO-81-61-L 81mm mortar (LMLB).

Designation	Mortier de 81mm Modèle 44
Manufacturer	Societe Nouvelle des Etablissèmènts Brandt, Paris
Dates in Irish service	1936-1981 (?)
Number in Irish service	48 (?)
Caliber	3.2 inches (81.4mm)
Crew	5
Weight (complete)	127 lbs (57.6 kg) approximate
Weight, barrel	40 lbs (18.1 kg)
Weight, bipod	40.8 lbs (18.5 kg)
Weight, base plate	45.2 lbs (20.5 kg)
Barrel length	40 inches (1016mm)
Angle of fire	+45° to +80°
Traverse	8° to 12°
Ammunition types	Steel bomb A.31 and steel bomb G.31, both filled with TNT;practice bomb filled with sand; practice bomb containing 6 oz.black powder; manouvre bomb; smoke bomb
Shell weight	7.26 lbs (3.3 kg) standard bomb; 14.5 lbs (6.6 kg) heavy bomb
Muzzle velocity	NA
Rate of fire	18 rounds per minute sustained
Range (maximum)	3400 yards (3109 meters) standard bomb; 1275 yards (1166 meters) heavy bomb

Hotchkiss Brandt MO-81-61-L 81mm mortar (LMLB)

The 81mm LMLB (Light Mortar, Long Barrel) mortar is the heaviest piece of firepower organic to the infantry battalion. In 1981 it replaced the older shorter-barrel 81mm Mk 44. Fitted with a C2A1 sight, and used in conjunction with a laser rangefinder, it is capable of hitting targets over five kilometers distant. The 81mm mortar is issued to all infantry battalions (both Regular and Reserve). Each battalion has a mortar platoon with three sections, and each section has two mortars, for a total of six mortars in the battalion and eighteen in the brigade. It is the battalion commander's organic artillery, and is under his direct control, providing close and immediate support as necessary. Some of the LMLB mortars are assigned to artillery units on an *ad hoc* basis.

Training with the 81mm Hotchkiss-Brandt LMLB mortar at Cemetery Hill in 1994. The 81mm LMLB replaced the short-barrelled Brandt M44 in 1981. Its longer barrel increases the range capability considerably. (62 Reserve Artillery Regiment, via Capt Ian Stewart)

An 81mm Hotchkiss-Brandt LMLB during training in 1994 at the 62nd Field Artillery Regiment winter camp. Although issued to infantry units, the LMLB occasionally is issued to artillery units, hence the reason for training with the weapon by reserve artillery units. (62 Reserve Artillery Regiment, via Capt Ian Stewart)

Designation	81mm MO-81-61-L
Manufacturer	Hotchkiss Brandt, France
Dates in Irish service	1981 -
Number in Irish service	72 (?)
Caliber	3.2 inches (81.4mm)
Crew	5
Weight (complete)	91.5 lbs (41.5 kg)
Weight, barrel	32 lbs (14.5 kg)
Weight, bipod	27 lbs (12.2 kg)
Weight, base plate	32.5 lbs (14.8 kg)
Barrel length	1450mm
Ammunition types	HE, smoke, illumination
Shell weight	7.26 lbs (3.3 kg)
Muzzle velocity	825 fps (251 m/s) approximate
Rate of fire	12 rounds per minute normal
Range (maximum)	109-6425 yards (100-5875 meters)

Brandt AM 50 120mm mortar

The 120mm Brandt AM 50 entered Irish service in 1953 and served with the Heavy Mortar Batteries in the field artillery regiments, generally as FCA batteries. The first actual firing of the AM 50 in Ireland was conducted in May 1954 by the 8th Field Artillery Battery near Oughterard, County Galway. Artillery mortar Troops equipped with the AM 50 were deployed to the Congo, Cyprus, and Lebanon, where they provided illumination fire missions in support of UNIFIL operations. The AM 50 was mounted on an integral wheeled carriage that normally was towed by the Land Rover until the 1990s, following which the mortar was towed by the newer Nissan Patrol

A layer from the 6th Field Artillery Regiment readying a 120mm Brandt for fire in 1998. The layer is wearing active ear protection. (62 Reserve Artillery Regiment, via Capt Ian Stewart)

A 6th Field Artillery Regiment loader dropping a HE round into the barrel of the AM 50 mortar, at Stranahealy gun position, 1994. The layer has a 6th Field Artillery Regiment cloth insignia on his arm brassard; the stars below the regimental badge denote a 3 Star Gunner. (62 Reserve Artillery Regiment, via Capt Ian Stewart)

vehicles. Due to the nature of the tubeless tires on the mortar carriage, it could be towed only for short distances at slow speeds (such as from Coolmoney Camp to the Glen of Imaal firing ranges), but for long distances the mortar had to be broken down into component parts and loaded onto Bedford or Leyland DAF trucks. The AM 50 was last fired in 2002, and was withdrawn from service in 2004, although some were maintained for training purposes. It was not until 2008 that the Irish Army selected and took delivery of a replacement mortar, the 120mm RUAG M87.

A 120mm Brandt AM 50 mortar in firing order with the Irish UNIFIL contingent, 1980. A large hole has been dug for the base plate, lowering the profile of the mortar. The crew member is holding an illumination round; empty ammunition boxes are in evidence to the side of the mortar. (Martin Forde via Roy Kinsella)

The gunner pulling the lanyard to fire the Brandt 120mm mortar. The AM 50 could be fired either by gravity or by a firing lever. The preferred method of firing was by use of the firing lever, actuated by a lanyard. (62 Reserve Artillery Regiment, via Capt Ian Stewart)

An FCA mortar team loading a 120mm round into the Brandt mortar. The crew are wearing the 1944 pattern helmet that was used until the late 1990s; the helmet was very unpopular due to most lacking a liner, using an elasticized strap instead, which caused the helmet to bounce uncomfortably while running. The webbing is the 58 pattern webbing, which was not very popular with the troops. The uniform itself consists of a shirt and fatigue trousers, used until introduction of the DPMs to the Reserves (Irish Defence Forces via Colin Stone)

A 120mm Brandt set up for firing at Stranahealy in 1994. The mortar remains on its wheeled carriage, with the base plate firmly planted on the ground. The blue nylon cord serves as a lanyard for firing the mortar. Several sheep, barely visible in the background, seem unconcerned by the activity. (62 Reserve Artillery Regiment, via Capt Ian Stewart)

The No. 2 of the mortar detachment (the loader), loading a high explosive round into a 120mm Brandt, 1998. (62 Reserve Artillery Regiment, via Capt Ian Stewart)

Designation	Mortier Brandt de 120mm Type AM 50
Manufacturer	Societe Nouvelle des Etablissèmènts Brandt, Paris
Dates in Irish service	1953 – 2004
Number in Irish service	72
Caliber	4.72 inches (120mm)
Crew	4
Weight (complete)	955 lbs (434 kg) on wheels; 585 lbs (260 kg) on bipod mounting
Weight, barrel	155 lbs (70 kg)
Weight, bipod	185 lbs (84 kg)
Weight, base plate	234 lbs (106 kg)
Carriage and accessories	405 lbs (184 kg)
Barrel length	69 inches (1750mm)
Ammunition types	HE, smoke (charges 0-7)
Shell weight	28.66 lbs (13 kg)
Muzzle velocity	390 fps to 1086 fps (119 m/s to 331 m/s) depending on charge
Rate of fire	NA
Range (maximum)	7376 yards (6745m) at charge 7

RUAG M87 120mm mortar

The RUAG 120mm M87 mortar replaced the earlier Brandt AM 50 120mm mortar. Upon withdrawal of the Brandt from service in 2004, the Army began to search for a replacement, eventually settling on the purchase of 95 second-user RUAG 120mm mortars from Switzerland. Deliveries of the RUAG to Ireland began in late 2008. The mortars were all reconditioned and were in immaculate condition. The Defence Forces also bought a stock of appropriate spare parts as well as ammunition as part of the acquisition package. The RUAG is particularly attractive as it can be fitted with

The 120mm RUAG mortar set up to fire, 62 RAR camp, July 2009. The mortar has bedded itself firmly into the ground after firing a few rounds. (Colin Stone)

Preparing to fire the RUAG on 9 July, 2009, during the 62 RAR practice shoots at the Glen of Imaal. The base plate is well dug in, and in the foreground is the container for eight 120mm rounds. (Colin Stone)

Firing the new 120mm RUAG mortar, Glen of Imaal. In 2008 the Irish Army acquired a large number of reconditioned RUAG mortars from Switzerland, enabling them to completely replace the ageing 25-pdrs in the Reserve Artillery Regiments. (Lt Col Edmond Fogarty)

A mortar round from a 120mm RUAG in flight. The gunner has fired the mortar by means of a lanyard, as was common practice with the 120mm mortar previously in Irish service, the Hotchkiss-Brandt AM 50. (Irish Defence Forces)

In addition to replacing the 25-pdrs in Reserve artillery regiments, the RUAG also equips one battery in each of the Irish Army's Regular artillery regiments. A flame halo is frequently observed when firing the RUAG mortar. (Irish Defence Forces) Reserve Artillery Regiment, via Capt Ian Stewart)

A 120mm RUAG mortar on its carriage, in the travel position. In addition to the mortar and its base plate, the trailer carriage carries a six rounds of 120mm ammunition, as well as pioneer tools and other equipment and accoutrements associated with the mortar. (Colin Stone)

an insert that allows firing 81mm rounds, leading to cost savings in training as well as providing operational flexibility. A properly trained crew is able to bring one of the mortars into action in three minutes.

The weapon is a towed weapon that comes with its own trailer unit that carries all the equipment necessary to deploy the mortar in a tactical situation. There is a six bay storage unit for ammunition on the trailer. The mortar is removed from the trailer and set up as a conventional mortar with bipod and base plate for firing.

Each PDF artillery regiment has one battery (for a total of three PDF batteries) equipped with the RUAG, and each RDF artillery regiment has three batteries (for a total of nine RDF batteries) equipped with the RUAG. The scale of issue is six mortars per battery, for a total of 72 RUAG mortars issued to units, leaving a substantial number as replacement mortars and for training use.

Designation	RUAG M87 120/81mm mortar
Manufacturer	RUAG Land Systems, Thun, Switzerland
Dates in Irish service	2008 – present
Number in Irish service	95
Caliber	4.72 inches (120mm)
Crew	6 (sergeant, corporal, 4 gunners, one of whom is a driver)
Weight (complete)	1650 lbs (722 kg)
Weight, barrel	187 lbs (85 kg)
Weight, bipod	110 lbs (50 kg)
Weight, base plate	206 lbs (94 kg)
Carriage and accessories	673 lbs (306 kg)
Barrel length	177mm
Ammunition types	HE, illumination, smoke (charges 0-7)
Shell weight	31.75 lbs (14.4 kg)
Muzzle velocity	NA
Rate of fire	6 rounds per minute
Range (maximum)	8968-10061 yards (8200-9200 meters); planning range for operational use 8202 yards (7500 meters)

A dramatic image of the 120mm RUAG firing at night, creating a rather impressive flame signature. As of early 2012, the RUAG represents both the newest and the most widely issued weapon in the inventory of the Irish Artillery Corps. (Lt Col Edmond Fogarty)

An assembled M87 RUAG tube and base plate, with its transport carriage behind it. Further in the background is a 25-pdr gun. This equipment is assigned to 2 FAR, Dublin. (Colin Stone)

Anti-Aircraft Artillery

QF 40mm Mk III (40mm Bofors L/60) anti-aircraft gun

The 40mm Swedish Bofors L/60 is somewhat of an iconic weapon amongst anti-aircraft guns. During World War II it was the most widely used gun of its type, and was used in every theater of the war by major and minor combatants on both sides. The design of the gun was drawn up in 1928, and the first gun was produced in Sweden in 1930. The gun was produced as a single mounting for land service, and as a twin mount for naval service. By 1939 the gun was a huge success, having been exported to or license produced in at least eighteen countries. Britain adopted the gun, but made numerous detail changes resulting in several different Marks of the gun, and at least half a dozen different Marks of carriage, but all were referred to generically simply as the Bofors. The most widely produced version was the QF 40mm Mk III. The L/60 served throughout the war, and was later upgraded to the L/70. The QF 40mm entered Irish service in 1939; the first four L/60s in Irish service, delivered in 1939, were Swedish-manufactured guns, but the rest of the L/60s were obtained from Britain and were the so-called Bristol Bofors, manufactured by Bristol Aircraft in England. In service with the Irish Army from 1939 through 2001, the L/60 served longer than any other gun type in the Irish Army. In 2001, the L/60 guns were replaced by the improved 40mm Bofors L/70 guns.

A 40mm Bofors L/60 at Cathal Brugha Barracks, 31 August 1998, at a display marking the 50th anniversary of the "End of The Emergency." The Bofors L/60 was a very reliable system, and served in the Irish Army for more than sixty years, an enviable record for any gun. (Denis McCarthy)

A 40mm Bofors L/60 at Cathal Brugha Barracks in August 1998. The Swedish-designed Bofors L/60 was an extremely effective weapon and was widely used by both Allied and Axis forces during WW II. (Denis McCarthy)

A Bofors L60 40mm gun firing in the Glen of Imaal, date unknown. Due to range limitations, as well as the lack of airborne targets, the gun is firing in a flat trajectory, with the gun depressed in the horizontal position. (Colin Stone)

An L/60 in Cathal Brugha Square, set up in a static position. The stabilizing legs are extended and leveled, and the wheels have been removed from the carriage. (Colin Stone)

A Bofors L/60 40mm anti-aircraft gun at a show, 4 May 1983. The initial lot of L/60 guns acquired by Ireland were manufactured in Sweden, but subsequent guns provided by Britain were of British production. (Denis McCarthy)

Designation	QF 40mm Mk III
Manufacturer	Bofors, Sweden, and Bristol Aircraft
Dates in Irish service	1939-2001
Number in Irish service	24
Caliber	40mm (1.6 inches)
Crew	7 (two NCOs, five gunners)
Elevation	-5° to +90°
Traverse	360°
Weight	5418 lbs (2458 kg)
Length	256 inches (6502mm) traveling length
Width	73.5 inches (1867mm)
Height	77.25 inches (1962mm)
Carriage	Mobile 4-wheel carriage with stabilizer arms
Barrel length	88 inches (2244mm)
Breech	Vertical sliding breech
Recuperator	NA
Ammunition types	HE, HE-I, HE-T
Shell weight	1.96 lbs (0.89 kg)
Muzzle velocity	2890 fps (881 m/s)
Rate of fire	60-90 rounds per minute (practical); 120 rounds per minute (cyclic)
Range (maximum)	23,600 feet (7193 meters)

Bofors 40mm L/70 anti-aircraft gun

The 40mm Bofors L/70 anti-aircraft gun is an improved version of the well-tested and highly successful 40mm L/60 gun. Following World War II, in order to cope with the threat posed by the advent of faster aircraft, it was apparent that an improved gun was needed that could fire at a faster rate and at longer range than the L/60. The resulting L/70, with its longer barrel firing a lighter shell with improved fuzing, with a cyclic rate of fire almost three times as fast as that of the L/60 (330 rounds per minute compared to the L/60's 120 rounds per minute) satisfied those requirements. The carriage is similar to that of the earlier Bofors L/60 gun, but is stronger and has an integral power unit that provides electro-hydraulic elevating and traversing for the gun. The first version was produced in Sweden in 1947, and in November 1953 it was accepted as the NATO standard anti-aircraft gun. In 2001 Ireland acquired thirty Bofors L/70 guns along with eight Signaal Flycatcher radar target acquisition and fire control systems; in characteristically parsimonious fashion, the equipment was acquired second-hand from the Netherlands. Twenty-four of the L/70s replaced an equal number of the now-obsolete Bofors L/60 guns that constituted the tube artillery equipment of the Air Defence Regiment's four batteries.

Although it bears many external similarities to the earlier L/60, the EL/70 is a somewhat larger and considerably heavier system. The EL/70 has a large shield absent on the L/60, as well as a prominent ammunition feed system. (Colin Stone)

An EL/70 in camouflage paint scheme at College Green, Dublin, 16 April, 2006. The EL/70 systems in Irish service were purchased from The Netherlands in 2001 and replaced the older L/60 guns as the primary air defence weapon in the inventory. (Sean O'Sullivan)

A Bofors EL/70 at Cathal Brugha Barracks, 5 May 2011. The L/70 is an enhanced version of the L/60 and is a much more capable system, especially in terms of rate of fire and range. (Colin Stone)

A lineup of EL/70s covered by tarpaulins parked at Curragh Camp. The photograph may date to 2001 when the guns were first delivered, as the last gun in line still has a small Dutch flag painted underneath the right rear taillight and reflector. (Sean O'Sullivan)

Designation	40mm Bofors L/70 anti-aircraft gun
Manufacturer	Bofors Sweden
Dates in Irish service	2001 -
Number in Irish service	30
Caliber	40mm (1.6 inches)
Crew	7 (two NCOs, five gunners)
Elevation	-20° to +80°
Traverse	360°
Weight	11,353 lbs (5150 kg)
Length	294 inches (7468mm)
Width	91 inches (2311mm)
Height	110 inches (2794mm)
Carriage	Mobile 4 wheel carriage with stabilizer arms
Barrel length	110.2 inches (2800mm)
Breech	Vertical sliding breech
Recuperator	NA
Ammunition types	HE-T, HE-AP-T, solid shot
Shell weight	1.9 lbs (0.87 kg)
Muzzle velocity	3320 fps (1012 m/s)
Rate of fire	330 rounds per minute
Range (maximum)	41,010 feet (12,500 meters)

QF 3 inch 20 CWT anti-aircraft gun

The QF 3 inch 20 cwt anti-aircraft gun was a 1914 design based on a pre-WW I Vickers 76mm naval gun. Its designation referred to its 3 inch (76mm) caliber as well as to its weight (20 hundredweight, or one ton), to distinguish it from several other 3 inch guns in use by the British at the time. It became the standard British home defence anti-aircraft gun used against German airships and bombers, as well as being used on British warships and submarines. The guns also served Britain in WW II until they could be replaced by the more modern 40mm Bofors and 3.7 inch guns. In March 1928 Ireland received four 3 inch 20 cwt AA guns (two 2-wheeled and two 4-wheeled), but it was not until 1931 that they were formally organised as Ireland's first anti-aircraft battery. They remained Ireland's only anti-aircraft weapons until the arrival of the first 40mm Bofors guns in 1939. All four reportedly were still carried in the inventory in 1949 even though they had been retired from active service some three or four years earlier, but ultimate disposition is unknown.

A 3 inch 20 cwt gun on an exercise in 1941. The gun has been raised on its stabilizer jacks, and the crew appear to be ready to fire. Note the rifles stacked in regulation fashion to the left of the gun. (Irish Defence Forces)

A pair of 3 inch 20 cwt guns firing at Gormanston in 1943. The 3 inch 20 cwt guns were the earliest anti-aircraft guns in the Irish inventory, having been acquired in 1928. The four guns of this type served during The Emergency, integrated into the air defence system around Dublin. (Lt Col Edmond Fogarty)

Training with the 3 inch 20 cwt guns at Gormanston in 1943. By this time Ireland had received enough of the more modern 3.7 inch heavy anti-aircraft guns to ease dependence on the ageing 3 inch 20 cwt guns. (Lt Col Edmond Fogarty)

Designation	QF 3 inch 20 cwt anti-aircraft gun, Mk I – Mk IV
Manufacturer	Vickers
Dates in Irish service	1928 – 1945/46
Number in Irish service	4
Caliber	3 inch (76mm)
Crew	11
Elevation	-10° to +90°
Traverse	360°
Weight	13,420 lbs (6087 kg)
Length	294 inches (7468mm)
Width	91 inches (2311mm)
Height	110 inches (2794mm)
Carriage	2-wheeled carriage; wheels could be removed and cruciform platform deployed for stability. Also made with 4-wheel carriage.
Barrel length	140 inches (3556mm)
Breech	Varied by Mark; semi-automatic sliding block or single-motion Asbury breech with Welin screw
Recuperator	Hydro-spring, constant
Ammunition types	HE
Shell weight	16 lbs (7.26 kg)
Muzzle velocity	2000 fps (610 m/s)
Rate of fire	16-18 rounds per minute
Range (maximum)	22,000 feet (6706 meters)

QF 3.7 inch anti-aircraft gun

In 1933 the British General Staff issued a specification for a 3.7 inch anti-aircraft gun. In January 1934 Vickers submitted a design, and completed a pilot model in April 1936. The first guns came off the production line in January 1938. The 3.7 inch QF became Britain's primary heavy anti-aircraft gun during WW II, serving throughout the war. There were two basic versions of the gun, one on a four-wheeled mobile mount, and the other in a static mounting with a large counterweight arm above the breech. Although it was quite similar to the more famous German 88mm Flak guns, the British rarely employed it as an anti-tank weapon as the Germans did so effectively as a matter of course. Ireland received six of the 3.7-inch guns in 1940, and an additional twelve in 1942. The six original guns were deployed in the eastern suburbs of Dublin during The Emergency and took part in defending Dublin against the German bombing of the city on the night of 23 April 1941. The 3.7 inch continued in service with the No. 3 Mk 7 radar which provided much greater accuracy than previously. The mechanical predictor was fitted with a "radar head" which permitted it to receive bearing, range and altitude target data from the radar, which in turn was converted into bearing,

A pair of 3.7 inch mobile anti-aircraft guns at Ringsend Park, Dublin, in 1941. These guns were the heaviest guns in the anti-aircraft defence arc located to the east of Dublin, close to the harbour. (MA, via Lt Col Edmond Fogarty)

A posed photograph of a 3.7 inch AA gun, date unknown. The 3.7 inch gun was used exclusively as an anti-aircraft weapon by the Irish Army, but could be used as a dual-purpose gun in an anti-tank role with proper ammunition. Its performance was comparable to that of the iconic German 88mm gun. (MA, via Lt Col Edmond Fogarty)

A 3.7 HAA at Curragh Camp, 27 May 2009. The stabilizers have been removed, but the gun still presents an imposing appearance resting on the concrete plinth. This is the sole surviving example of the eighteen guns of this type in Irish service, and previously was displayed at McGee Baracks, Kildare. (Ralph Riccio)

Dublin schoolboys enjoying a hands-on experience with a 3.7 inch anti-aircraft gun, whose proportions seem even more massive next to the slight stature of the boys. Photograph taken at the Military Tattoo in Ballsbridge, 1945. (MA, via Lt Col Edmond Fogarty)

range and fuze setting for use by the guns. Four remained in the inventory in 1949, and the last shoot with these guns took place in 1959. They were boarded in the early 1970s, and all except one, which has been restored and is on static display at Curragh Camp, were disposed of as scrap metal. Some 70-80,000 rounds of ammunition were also disposed of along with the scrapping of the guns.

The personnel standing near the 3.7 AA gun provide a reference as to the size of the rather large gun. (Colin Stone)

Designation	QF 3.7 inch anti-aircraft gun
Manufacturer	Vickers
Dates in Irish service	1939 – 1959
Number in Irish service	18
Caliber	3.7 inches (94mm)
Crew	7
Elevation	-5° to + 80°
Traverse	360°
Weight	20,451 lbs (9276 kg)
Length	342 inches (8687mm)
Width	96 inches (2438mm)
Height	98.5 inches (2502mm)
Carriage	Mobile mount on 4-wheel carriage, or static mount
Barrel length	185 inches (4699mm)
Breech	Horizontal sliding breech
Recuperator	Compressed air; oil buffer
Ammunition types	HE, AP, shrapnel
Shell weight	28.56 lbs (12.95 kg)
Muzzle velocity	2600 fps (792 m/s)
Rate of fire	10-20 rounds per minute
Range (maximum)	32,000 feet (9754 meters)

RBS 70 Man-portable air defense missile system

The RBS 70 man-portable air defense system (MANPADS) is a Swedish designed and developed short-range air defense missile system. The system was introduced into the Irish air defence inventory in 1980, and represents Ireland's only surface-to-air missile system. In 2008 an unspecified number of RBS 70 were added to the inventory. Hallmarks of the RBS 70 system are its low cost, ease of use, extreme mobility, ease of maintenance, and effectiveness. The system is an optical beam riding system. A local combat control terminal, about the size of a laptop computer, provides target position information to the operator; upon target acquisition, the operator turns off the safety feature, which then switches on the laser and sends out an IFF signal. Upon being fired, the missile rides the laser beam to the target, adjusting position if and as operator changes the beam's position. Exhaust is vented in the missile's midsection; the laser beamriding system is mounted in the tail, making it very difficult to jam. If the missile is guided to within 30 meters of the target, kill rate assurance is 95%.

The RBS 70 gunner seated on the very compact launch unit. The mast and dish of the supporting acquisition and tracking radar are visible in the background. (Irish Defence Forces)

An RBS 70 on its tripod, with a rudimentary camouflage covering, taken at Mullingar on Artillery Day 1998. The Nissan Patrol RD28 in the background is towing a Clancy Barracks built Irish pattern trailer used to transport the RBS 70 system. (Colin Stone)

The RBS 70 launch unit up close. The sponge rubber end caps provide protection to the launch tube while being transported. The system is compact and easily transportable. Although the system has been in the Irish inventory for some thirty years, it remains an effective, reliable weapon well suited to Irish requirements. (Irish Defence Forces)

Designation	**RBS 70 Man-portable air defense missile system**
Manufacturer	Saab Bofors Dynamics, Karlskoga, Sweden
Dates in Irish service	1980 – present
Number in Irish service	6 confirmed, plus additional unspecified number
Diameter	3.7 inches (94mm)
Crew	3 (one NCO, two gunners)
Elevation	NA
Traverse	360°
Total system weight	182 lbs (82.5 kg)
Length (missile)	52 inches (1320mm)
Engine	Booster and sustainer with smokeless solid propellant
Warhead	2.4 lbs (1.1 kg)
Velocity	Mach 1.6 to Mach 2 (approximately 1225 mph/1970km/h to 1530 mph/ 2460 km/h)
Muzzle velocity	2600 fps (792 m/s)
Rate of fire	NA
Range (maximum)	4.3 miles (7 km); altitude 2.5 miles (4 km)

Coast Artillery

QF 12 pounder 12 cwt coast defence gun

The QF 12 pounder was introduced as a naval gun in 1894 and was used until the middle of the 20th century on Royal Navy warships. Many of these guns were adapted as coast artillery guns and mounted on pedestals secured in protected emplacements to protect harbors against attack by small fast vessels such as torpedo boats. As coastal artillery, the guns were traversed manually by the gunlayer, who stood on the left side of the piece as he aimed. He elevated the piece with his left hand, and firing with his right hand. Eight 12 pounders were located at Fort Bere Haven, and four were located at Fort Templebreedy, guarding Cork Harbour. These coastal defense guns served in the UK until the 1950s, as they did in Ireland possibly until the 1960s. As of 2012, four are still maintained at Fort Davis in Cork Harbour for saluting purposes, and another four similarly are maintained at Dun Laoghaire for the same purpose.

Four 12-pdr guns on the East Pier at Dun Laoghaire. Date of photograph is unknown, but as of early 2012 the guns were still in place, for occasional use as salute guns. (Dun Laoghaire Historical Society).

12-pdr salute guns at Dun Laoghaire on 17 September 2009. Despite the fact that the guns are covered by protective tarpaulins, periodic maintenance remains necessary, especially due to the close proximity to salt water. (Roy Kinsella)

A 12-pdr at Newhaven Fort, Sussex, England. This 12-pdr was obtained from Ireland and cosmetically refurbished. Although the gun barrel itself seems to be free of corrosion, the breech, cradle, and gun mount show evidence of significant rusting underneath the paint. (Ed Tyhurst, via Philip Baldock)

The 12-pdr battery at Dun Laoghaire saluting an Argentine Navy training vessel in 1998 or 1999. The Sergeant in the foreground is wearing a shoulder patch that indicates service with the Irish Ranger Wing. All of the artillery personnel, from 2nd FAR, are wearing the white lanyard. The instrument in the center of the picture is a depression range finder; the hole in the metal plate contains an optic which was calibrated and allowed a gunner to calculate the rise or fall of the tide. (RSM Mick Dillon via Colin Stone)

A 12-pdr at Fortyfoot in the early 1940s. A battery of 12-pdrs was attached to the air defence establishment during the early part of The Emergency, and guns such as this dotted the coast to the southeast of Dublin. The soldier in the foreground is wearing a Volunteer jacket, with six buttons in the back. (Irish Defence Forces)

A 2009 photograph of the stonework base upon which the 12-pdr at Fortyfoot was mounted during The Emergency. The wall and base are now located in the garden of a private home. (Roy Kinsella)

One of the Fort Davis 12-pdr guns ready to fire a salute. Three salute rounds can be seen to the left of the gun. (Irish Defence Forces)

Two of the four 12-pdr guns at Fort Davis in Cork Harbour, normally kept covered with tarpaulins. The guns are facing Fort Camden on the opposite shore, and were sited to provide interlocking fire with the Fort Camden guns. (Colin Stone)

Designation	QF 12 pounder 12 cwt coast defence gun
Manufacturer	Elswick Ordnance Company; Vickers
Dates in Irish service	1938 – present (as saluting guns as of 2012)
Number in Irish service	16 (?)
Caliber	3 inches (76mm)
Crew	3 (?)
Elevation	+20° to -15°
Traverse	360 degrees
Weight	1124 lbs (510 kg)
Length	123.6 inches (3139mm)
Width	19 inches approximate
Height	43.5 inches
Carriage	Pedestal mount
Barrel length	120 inches (3048mm)
Breech	Single-motion screw
Recuperator	Hydro-spring
Ammunition types	HE
Shell weight	12.5 lbs (5.7 kg)
Muzzle velocity	2210 fps (674 m/s)
Rate of fire	15 rounds per minute
Range (maximum)	11,750 yards (10,744 meters)

QF 4.7 inch gun Mk IV

The QF 4.7 inch gun was designed and developed in the late 1880s and served with the navies of various countries (primarily the British Royal Navy). It was designed to replace the older BL 5 inch (127mm) naval guns and was optimized for firing modernized smokeless powder. They were also used as coast defence artillery, as well as being mounted on various wheeled carriages to provide the British Army with a long range gun. There were a number of different Marks of the gun, from Mk I through Mk IV. Four 4.7inch guns were inherited by Ireland with the turnover of the coast artillery forts in 1938. Two were located at Bere Island, and two at Fort Dunree. One example survives and is on display at Fort Dunree.

One of the original two 4.7 inch guns at Fort Dunree has survived and is on display there. The gun, serial number 1603, is a QF 4.7B Mk IV, and was built in 1902. (David Magee)

Designation	QF 4.7 inch gun
Manufacturer	Elswick Ordnance Company/Armstrong Whitworth; Vickers; Maxim
Dates in Irish service	1938 – 1940s (?)
Number in Irish service	4
Caliber	4.724 inches (120mm)
Crew	10
Elevation	NA
Traverse	NA
Weight	NA
Length	NA
Width	NA
Height	NA
Carriage	Pedestal mount
Barrel length	189 inches (4800mm)
Breech	Single-motion interrupted screw
Recuperator	NA
Ammunition types	Separate loading QF: AP, shrapnel, HE, common Lyddte, common pointed
Shell weight	45 lbs (20.4 kg) HE
Muzzle velocity	2150 fps (648 m/s)
Rate of fire	5 – 6 rounds per minute
Range (maximum)	12,000 yards (10,973 meters)

BL 6 inch Mk VII naval gun

The BL 6 inch Mk VII gun was a versatile gun, and was improved through a number of different Marks, from Mk I through the QFC Mk VII. The BL designation indicated a breech loading weapon in which the breech mechanism itself sealed the firing chamber. In the BL system, the projectile was loaded into the gun tube, followed by cloth bags containing the propellant charge (in contrast to QF guns in which the projectile and propellant formed a single shell, with the propellant encased in a brass casing). Originally developed as a British naval gun in 1899, it also served as a heavy field gun during WW I, as well as serving on various classes of Royal Navy ships (mainly on battleships and armoured cruisers). It was also used as a coast artillery gun throughout the United Kingdom, with about a dozen located at various coast defence forts in Ireland. At the time of the turnover of the Treaty port forts to Ireland in 1938, there were eight operational 6 inch guns in the

This 6 inch gun at Newhaven Fort, Sussex, England, was originally installed at Duncannon Fort, Co.Wexford, Ireland, but was returned to England in 1982. The gun was restored to operational condition prior to being mounted in its new location at Newhaven Fort. (Ed Tyhurst, via Philip Baldock)

One of the 6 inch coast defence guns at Fort Davis in July 1987. The guns at Fort Davis were in above-ground positions and have remained exposed to the elements since their decommissioning in the late 1960s or early 1970s. (Denis McCarthy)

Brass plate on the 6 inch gun at Newhaven Fort, indicating that it was emplaced at Duncannon Fort, Waterford, Eire, specifying the trunnion height as 55 feet, 0 inches above mean sea level at that location. The gun's actual location would have been in County Westmeath, downstream and across the River Suir from Waterford. (Ed Tyhurst, via Philip Baldock)

A 6 inch gun emplacement in Cork Harbor at Cork. This appears to be a gun position on Spike Island, with Curraghbinny and Fort Meagher in the center and left background. (photograph courtesy Irish Examiner)

forts; in 1942 two more were acquired from Britain and installed at a new fort, Fort Shannon. Nine, and possibly all ten, of the guns have survived, although some are deteriorating badly. One ex-Irish 6 inch gun has been restored and is located at Newhaven Fort, Sussex, England.

Designation	BL 6 inch Mk VII naval gun
Manufacturer	Vickers
Dates in Irish service	1938 – 1960s (?)
Number in Irish service	10
Caliber	6 inches (152.4mm)
Crew	9 (2 NCOs, 7 gunners)
Elevation	+ 20° to – 15°
Traverse	Unknown
Weight	16,875 lbs (7655 kg) gun and breech
Length	279 inches (7087mm)
Width	70 inches (1778mm) including shield
Height	118 inches (2997mm) to top of gun shield
Carriage	Mk II (pedestal)
Barrel length	269.5 inches (6845mm)
Breech	Welin single motion parallel breech screw
Recuperator	Spring
Ammunition types	Mark XXB HE; Mark XVIIIB shrapnel
Shell weight	99.25lbs (45 kg)
Muzzle velocity	2556 fps (779m/s) (new gun)
Rate of fire	8 rounds per minute
Range (maximum)	13,350 yards (12207 meters) on Mk II mount

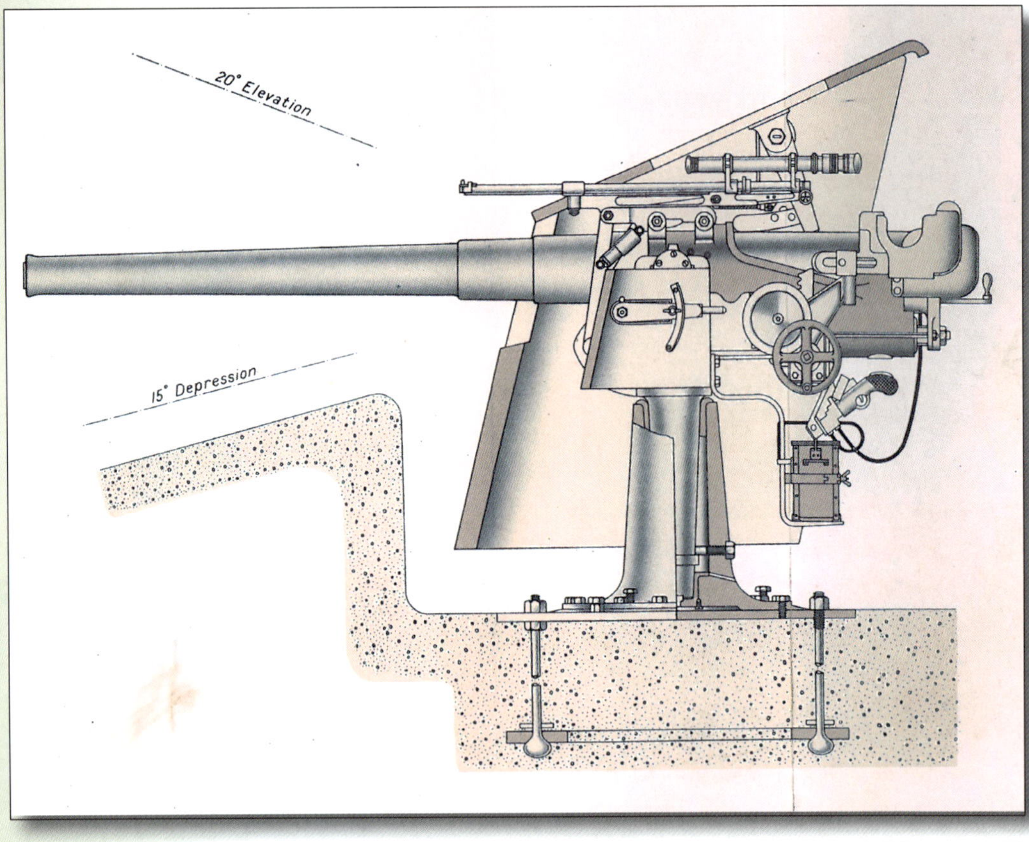

Schematic diagram of a 6 inch coast defence gun, indicating angles of elevation and depression. Throughout its lifetime, the 6 inch gun, developed in 1899 as a piece of naval ordnance, was also used as a field gun and as a coast defence gun. (Philip Baldock)

The 6 inch gun at Newhaven Fort being fired, after being refurbished, probably around 1983. The gun was formerly installed at a coast defence fort in Ireland. (Philip Baldock)

BL 9.2 inch gun Mark X

The BL 9.2 inch Mk X and the earlier Mk IX were naval and coast defence guns used by the British navy from 1899 to the 1950s. There were a total of five 9.2 inch guns in Ireland in 1938, with one at Fort Davis, and two at Templebreedy (guarding Cork Harbor), and two at Fort Lenan (Lough Swilly). The 9.2 inch gun at Fort Davis was transferred to Templebreedy to replace a defective gun there. The guns were mounted on Moncrief carriages, commonly known as "disappearing carriages," because as the gun was fired the recoil forces drove the gun backwards and downwards so that the gun itself disappeared below the level of its protective parapet. The gun was then able to be safely reloaded before being put in battery for the next shot. The forts and their guns were taken over by Ireland in 1938 when the British quit the Treaty Ports in accordance with the provisions of the 1922 Anglo-Irish Treaty, and were maintained in active service until the mid-1940s; by 1950 they were in a state of heavy preservation, and subsequently were scrapped.

Designation	9.2 inch/47 (23.4 cm) Mark X
Manufacturer	Elswick Ordnance Company; Vickers
Dates in Irish service	1938 – 1945/50 (?)
Number in Irish service	4
Caliber	9.2 inches (233.68mm)
Crew	NA
Elevation	35°
Traverse	NA
Weight	63,206 lbs (28,670 kg) including breech mechanism
Length	442.4 inches (11,237mm)
Width	NA
Height	NA
Carriage	Moncrief carriage ("disappearing" carriage)
Barrel length	429.3 inches (10,904mm)
Breech	Welin breech mechanism
Recuperator	NA
Ammunition types	APC, CPC, HE
Shell weight	380 lbs (172 kg)
Muzzle velocity	2872 fps (875 m/s)
Rate of fire	3 – 4 rounds per minute
Range (maximum)	Up to 36,700 yards (33,559 meters) with super charges

A 9.2 inch gun at Templebreedy, date unknown. The 9.2 inch gun was a piece of naval ordnance that was adopted for the coast defence role to protect coastal locations in England as well as Ireland. (photograph courtesy Irish Examiner)

Prime Movers

During the Civil War, the National Army seems to have relied exclusively on mechanical transport to tow, or carry, the 18-pdr field guns. Although Lancia armoured troop carriers are known to have been used to tow the guns during the fighting in Dublin in June and July 1922, as well as during the landings at Cork in August, both Crossley tenders and heavy commercial trucks as well were impressed to tow the field pieces or carry them portee-style. Specific details or descriptions as to how the guns were secured for firing from the truck beds is lacking, or if indeed the guns were actually fired from the trucks or if they were off-loaded and then manhandled into position. There is also at least one photograph of a Rolls-Royce armoured car (*Custom House*) towing an 18-pdr, but this does not appear to have been a widespread practice – indeed, a separate accompanying vehicle would have been needed to carry the artillery crew. There are only isolated accounts of the National Army employing horses to tow the guns, usually on an exceptional basis, and there is only scant evidence that guns were transported by rail. Rather, the norm was for the guns to accompany attacking columns along with the infantry and, almost invariably, armoured cars. The 18-pdrs were also a staple of General Emmet Dalton's seaborne expeditions against both the West and East coasts.

A Crossley tender belonging to the Army Transport Corps. Although during the Civil War the preference was to use Lancia personnel carriers to tow the 18-pdr field guns, there is at least one account of a Crossley tender being used to tow the 18-pdr, specifically during the attack on Waterford in July 1922.(Terry Ward)

Prime movers, or gun towers, in the Irish Army do not appear to follow a clear-cut pattern with respect to organisational assignment. Although PDF or Regular units have a certain number of trucks organic to the regiments, the old FCA or the newer Reserve Artillery Regiments essentially rely on trucks from the PDF/Regular regiments, or from the S&T Corps or logistical support units to provide the motive power for the guns. With the retirement of the 25-pdrs in July 2009 and the introduction of the 120mm RUAG mortar, combined with the introduction of the Mitsubish Pajero in mid-2009, movement of the artillery equipment in the field artillery organisation (now consisting only of the 105mm Light Guns and the RUAG) is considerably simplified, as the mortars can be towed by both the Nissan Patrol and Mitsubishi Pajero 4 x 4 vehicles, leaving only the 105mm guns in need of larger vehicles for movement.

Horses

From the mid 1920s until 1939 the Artillery Corps was largely equipped with horses used to tow the field guns then in service (the 18-pdrs, and to some extent the 3.7 inch pack howitzer). The horses on hand were the Irish Draught horse, which were large, strong animals well-suited to the task of towing the guns. An artillery battery had about 60 horses assigned; each gun was towed by a team of six horses. In early 1939 the first tests with modernized gun carriages were carried out; these tests were successful and paved the way for mechanization of the Artillery Corps. As a consequence, in March 1939 most of the horses were sold at public auction in Dublin, and the remainder parceled out to other units in the Army. By 1940 mechanization of the Artillery Corps had largely been accomplished by replacing the horses with Ford 2-3 ton trucks.

ACMAT

The French-produced ACMAT (*Ateliers de Construction Mècanique de L'Atlantique*) is a cross-country tactical military vehicle known for reliability, simplicity, and ruggedness. Originally targeted at the African and Asian markets, ACMAT trucks have been used by almost 50 countries, including Ireland. Layout of the ACMAT is conventional, with an engine compartment, an open driver's compartment to the rear of the engine, and a cargo bed in the rear. In December 1983, Ireland acquired the first of 16 ACMAT in the 4 x 4 VLRA (*Vèhicules de liaison, de reconnaissance, et d'appui*) TPK420 SM3 Light Reconnaissance and Support Vehicle variant. The ACMATs, fitted with right-hand drive, were purchased as artillery prime movers and served until about 1997. Although they were described by some drivers as excellent equipment in the off-road environment, and very powerful in the low end and in the 4 x 4 mode, they have also been described as difficult to drive on the road with limited vision due to the long front end. The ACMAT had no power steering, brakes were problematic, and the single rear axle tended to make them wander when cornering. Speed when towing the 25-pdr was kept to about 40 miles per hour. The canvas doors and lack of a heater made for very cold driving at times in the Irish weather.

A team of six horses pulling an 18-pdr and its limber during a ceremonial march past in Dublin in 1924. This gun and crew likely belonged to the 1st Field Battery stationed at McKee Barracks. Until 1939 Irish light draught horses were used to tow artillery pieces and their limbers, but between 1939 and 1940 the artillery carriages were updated with pneumatic tires, allowing the pieces to be towed by motor transport. (Irish Defence Forces)

A mounted artillery escort during the funeral of Michael Collins. Interestingly, the preponderance of horses were to be found in National Army artillery units, but not in cavalry units, which were motorised. Also of note is the fact that during the Civil War there was a decided preference for motor transport over horses to tow artillery pieces whenever possible. (Broe Collection)

An ACMAT with 25-pdr hitched for transport, Glen of Imaal Camp, 1996. The brown tubes between the gunners are hard cardboard packaging that contain the cartridge cases. The ammunition being held by the gunners is British origin, as denoted by its buff colour, and is being checked prior to firing. The small green box under the seat contains the sights for immediate use. An FN FAL service rifle is clearly visible. The crew, from 14th Batttery, consisted of No. 1 Sgt Leech, Coverer Cpl Dornan, No. 2 Gnr Devoy, No. 3 Gnr Stone, No. 4 Gnr Copley. To the right of the ACMAT is a Land Rover and crew belonging to the 19th Battery, 7th FAR equipped with the 120mm Brandt mortar. (Colin Stone)

A convoy of ACMATs towing 25-pdr guns in Phoenix Park. Obviously prepared for the ever-present possibility of rain in Ireland, both the cab and the body of the ACMAT are fitted with canvas covers. The 25-pdrs are minus their firing platforms, which allows a smooth gun drill and a quicker entrance and exit than would otherwise be possible. (RSM Mick Dillon via Colin Stone)

A 25-pdr hitched to an ACMAT VLRA prime mover. The ACMAT was a simple, reliable vehicle with excellent cross country capabilities well suited to towing the artillery in the Irish inventory. (Sean O'Sullivan)

ACMAT KZG 812 towing a 105mm L118 gun. The red wheel hubs used on all of the Cork ACMATs add a splash of colour. The gun's barrel is in the reversed position for travel. (Sean O'Sullivan)

An ACMAT towing an L 118 light gun during an exercise in the Glen of Imaal. A distinctive feature of the L118 is that its barrel can be reversed on the carriage for towing. (Military Archives)

An ACMAT VLRA still hitched to an L118 light gun whose barrel has just been rotated to the firing position, Glen of Imaal, 28 March, 1985. The ACMAT was able to cope well with the type of muddy field conditions pictured. (Denis McCarthy)

ACMAT VLRA command post version photographed on Cemetery Hill, Glen of Imaal. The flag on the right fender and the decal on the bumper mark the vehicle as belonging to 2nd FAR. During its lifetime, this particular vehicle had been painted in various liveries, including green, camouflage pattern, and gloss green, and possibly white while on UN duty. (Colin Stone)

An ACMAT VLRA TPK420 SM 3 with hard top cab and canvas tilt at Mullingar on Artillery Day, 1998, with a 105mm light gun hitched in the travel mode. The hard top cab affords the driver and passenger considerably more protection from the often harsh Irish elements than does the canvas enclosed version. (Colin Stone)

Designation	ACMAT
Manufacturer	*Ateliers de Construction Mècanique de L'Atlantique* (ACMAT),Saint-Nazaire, France
Dates in Irish service	1983-1997
Number in Irish service	16
Weight	9150-9480 lbs (4150-4300 kg)
Effective load	3307-5512 lbs (1500-2500 kg)
Length	232 inches (5900mm)
Width	83 inches (2100mm)
Height	60 inches (1500mm) (below tarp)
Engine	6-cylinder Perkins Diesel, 5.8 liter, 145 HP
Transmission	K 107 002; 4 forward, 1 reverse; transfer case allows 8 total speeds
Fuel capacity	2 x 120 liter fuel tanks (240 liters total) (52.8 Imperial gallons; 63.4 US gallons)
Wheelbase	142 inches (3600mm)
Track (front/rear)	69 inches (1760mm)/645 inches (1660mm)
Tire size	385/80 R20
Maximum speed (road)	56 mph (90 km/h)
Maximum range	373-621 miles (600-1000 km)
Maximum gradient	65%
Slope	30%
Ground clearance	12 inches (300mm)
Fording depth	35 inches (900mm)
Special notes	Air transportable by C-130 or C-160

Five ACMAT VLRAs lined up in front of their 25-pdrs at McKee Barracks, Dublin. Other photographs of this same grouping of guns and vehicles show all six guns to be present. (RSM Mick Dillon via Colin Stone)

AEC Matador

The British AEC Matador was a 4 x 4 artillery tractor developed primarily to tow medium artillery pieces. It was originally developed from a 4 x 4 designed by Charles Cleaver and produced by Hardy Motors and FWD in the early 1930s. Hardy Motors was absorbed by Associated Equipment Company AEC) in 1932, and development work continued on the vehicle. The first diesel engine version was delivered to the British Army in November 1939. Some 10,000 were built and used by British and Commonwealth forces during WW II. In Irish service, the first use of Matadors was between 1950 and 1952, when Matadors were hired from Ireland's CIE transport group to tow the 3.7 inch AA guns from Dublin to the firing range at Gormanstown, County Meath. The Army, not surprisingly, did not view this as an entirely satisfactory arrangement, and in early 1953 the Army acquired eight AEC Matador Mk III chassis; cabs and GS bodies subsequently were fabricated for them by the Base Workshops at Clancy Barracks, Dublin. The registration numbers for the Matadors were ZL 1251-1258. The Matadors, based at McKee Barracks, were the first Diesel trucks in the Irish Army inventory, and served until 1980. In addition to their primary purpose as gun tractors, the Matadors were also used to haul ammunition and supplies. One of the Matadors is preserved at the National Transport Museum in Howth.

The National Transport Museum's restored Matador, ZL 1257. The Matador was acquired by the museum in 1980, in badly vandalized condition, and was restored in 1987. (National Transport Museum, via Michael Corcoran)

Designation	AEC Matador
Manufacturer	AEC (Associated Equipment Company), Southall, London, UK
Dates in Irish service	1953-1980
Number in Irish service	4
Weight (unloaded/loaded)	15,850 lbs (7132 kg)/24,304 lbs (11,024 kg)
Effective load	22,400 lbs (10,160 kg)
Length	249 inches (6320mm)
Width	94 inches (2390mm)
Height	122 inches (3100mm)
Engine	AEC 6-cylinder Type 173 Diesel, 7581 cc (462 cu. In), 95 HP
Transmission	4-speed, with 2-speed transfer case
Fuel capacity	40 Imperial gallons (48 US gallons) (182 liters)
Wheelbase	151.5 inches (3848mm)
Track (front/rear)	NA
Tire size	13.50 x 20
Maximum speed (road)	36 mph (58 km/h)
Maximum range	357 miles (575 km)
Maximum gradient	NA
Slope	NA
Ground clearance	NA
Fording depth	NA
Special notes	Air transportable by C-130 or C-160

Austin K9

The Austin K9 4 x 4 25-cwt (2,800 pounds) trucks traced their origins back to the 1939 K-series trucks that were built for military use in WW II. After the war the K range continued with little change, but a more stylish 4 x 2 model designated the Loadstar emerged in 1950. A militarized version of the Loadstar evolved as the 4 x 4 K9, and in 1953 and 1954 the Irish Army took delivery of a total of six K9 25-cwt trucks, as a chassis and cab only. The wooden bodies were built and fitted at the Clancy Barracks Base Workshops. The K9s were used to tow the 25-pdr gun/howitzer. All six of the K9s were sold off in 1971.

Designation	Austin K9
Manufacturer	Austin Motor Co. Ltd., Longbridge, England
Dates in Irish service	1953-1971
Number in Irish service	6
Weight (unloaded/loaded)	NA
Effective load	NA
Length	206 inches (5232mm)
Width	87 inches (2210mm)
Height	94 inches (2388mm) to top of cab
Engine	Austin 6-cylinder gasoline, 3995cc
Transmission	Manual 4 forward, 1 reverse; transfer box
Fuel capacity	20 Imperial gallons (24 US gallons; 91 liters)
Wheelbase	137 inches (3480mm)
Track (front/rear)	71 inches (1803mm) front; 60 inches (1524mm) rear
Tire size	900 x 20
Maximum speed (road)	NA
Maximum range	NA
Maximum gradient	NA
Slope	NA
Ground clearance	NA
Fording depth	NA
Fording depth	35 inches (900mm)

17-pdr anti-tank guns being towed by Austin K9 trucks during the 1955 Easter parade on Lord Edward Street in Dublin. Only six of the K9s were acquired for service in the Irish Army, enough to tow one six-gun battery as needed. (Military Archives, via Denis McCarthy)

Bandvagn 206 (Bv 206) tracked vehicle

The Bv 206 is a light tracked vehicle (Bandvagn is Swedish for "tracked vehicle") originally developed by Hägglunds for the Swedish Army as an all-terrain vehicle specially suited for Swedish conditions. It has since been adopted for service by almost 40 countries. The Bv 206 is an articulated system consisting of two units, with all four tracks powered. Although normally used as a personnel carrier, it can be adapted to other roles, as is the case of the Bv 206 in Irish service. Although the number is unconfirmed, there are indications that five Bv 206 vehicles fitted with the Swedish Giraffe G50 radar system were delivered to the Irish Army in early 2009. These have supplemented the single Giraffe G40 radar mounted on the MAN 6 x 6 truck, in Irish service since 1986.

A Bv 206 cresting a rise during a training exercise in 2009. The Giraffe radar, here folded for travel, is extended upwards when operating to provide targeting and tracking data to supported gun and missile units. (Irish Defence Forces)

One of the Bv 206 articulated tracked vehicles acquired by the Irish Army in 2009. The vehicle mounts the Swedish G50 Giraffe radar system which provides target acquisition and fire control for the EL/70 and RBS 70 air defence systems. The tracks of both the front and rear units are powered, providing exceptional mobility. (Irish Defence Forces)

Designation	Bandvagn 206 (Bv 206)
Manufacturer	Hägglunds/BAE Land Systems
Dates in Irish service	2009 –
Number in Irish service	5 (?)
Weight (unloaded/loaded)	9921 lbs (4500 kg) (?)
Effective load	4938 lbs (2240 kg)
Length	272 inches (6900mm)
Width	74 inches (1870mm)
Height	94,48 inches (2400mm)
Engine	Ford V-6, 2800 cc
Transmission	MB W 4A-018 automatic
Fuel capacity	NA
Wheelbase	Tracked system
Track	48.4 inches (1230mm)
Maximum speed (road)	31 mph (50 km/h)
Maximum speed (water)	1.9 mph (3 km/h)
Maximum range	205 miles (330 km)
Maximum gradient	100%/45°
Slope	75% (35°)
Ground clearance	NA
Fording depth	Fully amphibious without additional equipment

A front left view of a BV 206; the Giraffe radar is in the folded position on the rear unit. (Colin Stone)

The rather complicated Giraffe radar array in the stowed position atop the rear element of the BV 206 system. Overall, the BV 206 with the radar is a compact system, with high mobility characteristics. (Colin Stone)

Bedford MFR2

The Bedford M-series trucks were developed in response to an early 1960s British requirement for a 4 x 4 4-tonne truck to replace the Bedford RL-series. Among the requirements were medium cross-country mobility, air transportability, and options for gasoline or Diesel engines. Three competing models were tested (Bedford, Austin, and Commer); the Bedford entry, based on the civilian 4 x 2 TK truck, was chosen. A contract was placed in 1968, and series production began in May 1970. The initial designation was MK, the K signifying a Diesel engine, but as widespread use of the gasoline engine became common, the vehicle became known as the M-type. The Irish Army acquired a number of Bedford MFR2 versions in the early 1970s, fitted with a Marshall drop-side metal cargo body. The Irish MFR2s were fitted with dual real wheels, in contrast to the MFRs in British service that were fitted with single rear wheels. The MFR2s were used as troop carriers, cargo vehicles, and as artillery prime movers. By 1987 there were 104 MFR2s in Irish service. According to Karl Martin, the vehicles left much to be desired. They were underpowered, had improper gear ratios, and suffered from reliability problems. Four of the Bedfords were converted to a Diesel configuration by installing the Bedford 330D six-cylinder Diesel engine, but their performance was not considered to be satisfactory enough to warrant further conversions, and accordingly the program was cancelled in 1988. The MFR2s continued in service until gradually replaced by the ACMATs that began entering service in 1983. Upon retirement from service, the Bedfords, which had been served beyond their useful life, were cut up; one source claims that Dieselized versions were exported.

A 3.7 inch mountain howitzer ready for towing by a Bedford MFR 2, Glen of Imaal, 1976. The MFR 2 was widely used as a gun tower, but did not have a reputation as being a particularly reliable vehicle. (62 Reserve Artillery Regiment, via Capt Ian Stewart)

Designation	Bedford MFR2
Manufacturer	Bedford Motors, Luton, England
Dates in Irish service	1973-1990s (?)
Number in Irish service	104
Weight (unloaded/loaded)	5 tons 4 cwt 35 lbs (11,763 lbs)
Effective load	NA
Length	257.5 inches (6541mm)
Width	98 inches (2489mm)
Height	128 inches (3251mm)
Engine	Bedford 6-cylinder gasoline, 4927 cc
Transmission	4-speed, plus 2-speed transfer case
Fuel capacity	33 Imperial gallons (40 US gallons; 150 liters)
Wheelbase	156 inches (3962mm)
Track	80.75 inches (2051mm) front; 80 inches (2032mm) rear
Tire size	9.00 x 20 (dual rear tires)
Maximum speed (road)	48 mph (77 km/h)
Maximum range	348 miles (560 km)
Maximum gradient	50%
Slope	30%
Ground clearance	NA
Fording depth	30 inches (760mm)
Fording depth	NA
Fording depth	35 inches (900mm)

Bedford MFR 2 8299 ZH being used for refuse collection during a strike by Dublin Corporation binmen in the early 1980s. Although far from hazardous or glamourous work, the service provided by the Army to the public on this and similar occasions filled a very real need. (Sean O'Sulivan via Colin Stone)

Bedford RL

Design and development of the R-series was begun in December 1950, with quantity production beginning in April 1952. Its design was based on that of the civilian 7-ton SLC chassis; the military chassis was designated the FV13100 series. The vehicle was offered in several versions, with a shorter wheelbase and with dual rear tires. It originally was rated as a 3 ton load carrier, but in 1968 was up-rated to 4 ton capacity. The Bedford RL 4 x 4 truck was the British Army's standard 3 ton medium truck from 1952 until 1969, when it began to be replaced by the M type. The Irish Army began to acquire the RL soon after its introduction into the British Army. Between 1953 and 1956, it took delivery of 128 RLs in cab and chassis form. Wooden bodies were built and added by the Clancy Barracks Base Workshops. An unspecified number of RL trucks also were bought in the 1960s. The RLs were assigned to infantry battalions and to the S & T Corps, as well as to artillery units where they were used to tow guns, replacing the WW II era Morris Quad Mk IIs.

The civilian origin of this Irish Army Bedford RL is evident from the rounded lines, quite in contrast to the generally angular appearance of most military trucks. The RL performed better in Irish service than did its successor, the Bedford MFR2. (Peter Leslie)

Bedford RL ZU 4939 being used as public transport while the state bus company was on strike. While certainly not as comfortable as a civilian bus, it nevertheless served the purpose during the transportation crisis. (Sean O'Sullivan, via Colin Stone)

Designation	Bedford RL
Manufacturer	Bedford Motors, Luton, England
Dates in Irish service	1953- 1980
Number in Irish service	128 plus
Weight	17,655 lbs (8025 kg)
Effective load	NA
Length	250 inches (6360mm)
Width	94 inches (2390mm)
Height	122 inches (3110mm)
Engine	Bedford 6-cylinder gasoline, 4927 cc, 110 HP
Transmission	4-speed, plus 2-speed transfer case
Fuel capacity	26 Imperial gallons (31 US gallons, 118 liters)
Wheelbase	156 inches (3962mm)
Track	73 inches (1854mm)
Tire size	11.00 – 20
Maximum speed (road)	NA
Maximum range	250 miles (402 km)
Maximum gradient	NA
Slope	NA
Ground clearance	NA
Fording depth	NA

Ford 1939 and 1940 Pattern V-8 2-3 ton trucks

British Ford 1939 and 1940 V-8 trucks were popular in Irish service. The Ford was in fact used as the basis for the Irish-designed and built Mk V and Mk VI armoured cars. The Fords were standard US pattern 1939 and 1940 commercial trucks, probably built at Dagenham, England, and assembled at the Ford factory in Cork, Ireland. It would appear from photographs that the trucks may have been purchased by the Irish Army as cab and chassis only, with the wooden drop-side bodies possibly being manufactured by Base Workshops. Because it was impractical to tow the 18-pdr field pieces at the speeds the Fords could achieve, until they were fitted with pneumatic wheels and modified for vehicle towing, the guns were loaded for transport portee-style onto the beds of the Fords. Presumably, the guns were manhandled onto the beds of the trucks with the assistance of loading ramps. The Fords were used from 1939 to 1960.

A Ford 1940 pattern truck sitting forlornly in a field in Gort, County Galway in the early 1980s. Its green paint has deteriorated badly, although its registration plate ZD 1648 remains well preserved. The dual rear wheel has been removed, presumably on both sides. The custom built wood body seems to have held up rather well. Note the tailgate which is in the down position. (Sean O'Sullivan)

Designation	Ford 1939 Pattern (Specifications are for Canadian Ford 3 ton)
Manufacturer	Ford, Cork, Ireland (assembled from components supplied by Ford, Dagenham, England)
Dates in Irish service	1939/1940 – 1960s
Number in Irish service	NA
Weight	NA
Effective load	NA
Length	212.95 inches (5409mm) front bumper to rear of frame
Width	85.56 inches (2173mm)
Height	79.39 inches (2017mm)
Engine	Ford flat head V-8, gasoline, 3621 cc (221 cu in) 85 HP (?)
Transmission	Heavy duty, 4 forward speeds
Fuel capacity	16.7 Imperial gallons (20 US gallons; 76 liters)
Wheelbase	134 inches (3404mm)
Track	57 inches (1448mm) front; 65 inches (1651mm) rear
Tire size	7.00 x 20, 8-ply
Maximum speed (road)	NA
Maximum range	NA
Maximum gradient	NA
Slope	NA
Ground clearance	NA
Fording depth	NA

Designation	Ford 1940 Pattern
Manufacturer	Ford, Cork, Ireland (assembled from components supplied by Ford, Dagenham, England)
Dates in Irish service	1939/1940 – 1960s
Number in Irish service	NA
Weight	NA
Effective load	NA
Length	205.32 inches (5215mm) front bumper to rear of frame
Width	86 inches (2184mm) approximate
Height	81.55 inches (2071mm)
Engine	Ford flat head V-8, gasoline, 3621 cc (221 cu in) 85 HP
Transmission	Heavy duty, 4 forward speeds
Fuel capacity	15.8 Imperial gallons (19 US gallons; 72 liters)
Wheelbase	134 inches (3404mm)
Track	73 inches (1854mm)
Tire size	7.00 x 20, 8-ply
Maximum speed (road)	NA
Maximum range	NA
Maximum gradient	NA
Slope	NA
Ground clearance	NA
Fording depth	NA

Lancia Jota and Triota

Large numbers of Lancia trucks and truck chassis were supplied by Italy to Britain during and following WW I. Two types of chassis were in the British inventory, the earlier Jota (or Iota) chassis, and the later, slightly larger Triota chassis. In 1920 British authorities arranged to have a significant number of armoured bodies manufactured in Ireland, built to British War Office pattern, to be fitted to the Lancia chassis in order to provide an armoured troop carrier to cope with the ongoing insurgency in Ireland. There were slight differences between the armoured bodies according to the chassis type. Those on the Jota chassis had the driver's cab sides that were no higher than the top of the troop compartment sides, and also had only a partial plate covering the front of the driver's compartment. The body on the Triota chassis differed in that it had a full angled plate covering the front of the driving compartment; the plate had a sliding vision port for the driver and a small vision or firing port to the left of the driver's port. The front plate was surmounted by a short plate that formed a brow in front of the open troop compartment, and the cab sides were slightly higher than the body sides. Most or all of the resulting vehicles were issued to the Royal Irish Constabulary (RIC). When Ireland became a Free State in 1922, among other vehicles handed over by the departing British were a number of Lancia armoured carriers (built on both types of chassis). There is doubt and controversy over precisely how many of the Lancias were handed over

A Lancia Triota personnel carrier providing cover to a Mk I 18-pdr at the corner of Henry and Sackville Street, ready to fire in the direction of "The Block". Lancias were frequently parked by the guns so as to provide a modicum of protection to the gun crew. Someone had been busy writing inscriptions with chalk on the side and front of the vehicle – the inscription on the side reads "We have no time for Truces", while that on the front plate reads "For Goodness Sake Aim Straight". (Kilmainham Goal Archive)

A Lancia Jota towing an 18-pdr field gun at Passage West, south of Cork City, during the National Army operation in August 1922 that led to the capture of Cork. (National Library of Ireland, HOGW 173)

to the Free State, the lowest (and in many respects the most believable) number being 64. There is no doubt, however, that the Lancias can claim the distinction of being the first artillery prime movers used by the National Army. On 28 June 1922, when National Army 18-pdrs (provided by the British garrison in Dublin) opened fire on Republican positions in the Four Courts building complex in Dublin, they had been towed there by Lancia armoured trucks. The Lancia's relatively low top speed made it suitable for towing the 18-pdr guns, with their wooden spoke wheels, although it is likely that speeds over extended distances were considerably slower. The Lancias were struck from the Army's vehicle inventory in 1933, but it is not clear for how long between 1922 and 1933 they continued to be used as prime movers for the 18-pdrs. For a period, until about 1939 or 1940, the Lancias were supplanted by horses to tow the guns.

Designation	Lancia Jota and Triota
Manufacturer	Lancia 1920-1921 truck chassis;armoured body built by Inchicore Workshops, Dublin
Dates in Irish service	1921-1932 (?)
Number in Irish service	64 (?)
Weight	9237 lbs (4190 kg)
Length	206 inches (5232mm)*
Width	102 inches (2590mm)*
Height	84 inches (2134mm)
Engine	Lancia Tipo 64 4-cylinder side-valve gasoline, 4940cc
Horsepower	70 BHP @2200 rpm
Transmission	Manual; 4-speed (?)
Fuel capacity	22 Imperial gallons (26.4 U.S. gallons; 10 liters)
Wheelbase	132 inches (3350mm)
Track	55 inches (1400mm)
Tire size	895 x 135 (twin rear)
Maximum speed (road)	37 mph (60 km/h)
Maximum range	249 miles (400 km)
Maximum gradient	NA
Slope	NA
Ground clearance	NA
Fording depth	NA

* Dimensions for Triota Mod. 1920-21 truck chassis

Land Rover

The Land Rover Series I originally was conceived as a light agricultural utility vehicle, and was based on the World War II US Willys jeep. The Land Rover was developed in 1947, and was put into production in 1948 and the vehicle was such a success that it has remained in production ever since, with successive improvements. Its longevity is a testament to its simplicity, ruggedness, and reliability. It has been produced in a huge number of versions, from the rather Spartan basic utility vehicle to upscale sport utility vehicles.

A short wheel base Land Rover, dating from about 1964. Various versions of the Land Rover were in Irish service as the principal light utility vehicle for more than thirty years. (Author's collection)

Between 1955 and 1986 the Irish Army took delivery of well over 500 Land Rovers in a number of different versions, beginning with a purchase of six 86-inch wheelbase Land Rovers in 1955, followed by another eleven in 1956. At least

One of the Series IIA short wheelbase Land Rovers purchased in 1965. The decal on the front fender identifies this vehicle as assigned to the 2nd Field Artillery Regiment, PDF, Eastern Command. The raised section on the center of the hood is the point where the spare tire normally is mounted. (Sean O'Sullivan)

A Land Rover Series III photographed in 2003. The paint shows considerable weathering; the paint on the right front wing has weathered to the point that the rustproof primer is showing through. (Sean O'Sullivan)

initially, all of the 86-inch wheelbase vehicles were assigned to the 4th Field Artillery Regiment and were used to tow the 120mm Brandt mortars. Although there was somewhat of a gap in acquisitions between 1958 and 1971, over the following years until 1986 the army bought varying numbers of the 88-inch, 107-inch, 109-inch and 110-inch wheelbase Land Rovers in varied configurations. In addition to towing the 120mm Brandt mortars, the Land Rovers also were used to carry the 81mm mortar and crew; rounding out the artillery applications, some Land Rovers were fitted with the Field Artillery Computer Equipment (FACE) system.

A 120mm Brandt mortar with the Irish UNIFIL contingent in Lebanon in 1980. In the background is a Land Rover in United Nations livery used to tow the mortar. (Martin Forde via Roy Kinsella)

Designation	Land Rover
Manufacturer	Land Rover, Solihull, England
Dates in Irish service	1954 -1990s
Number in Irish service	More than 550 of all versions combined
Weight	Varied according to model
Length	Varied according to model; 181 inches (4597mm) for 110
Width	61 inches (1549mm) to 70.5 inches (1791mm)
Height	73.5 inches (1867mm) to 84 inches (2134mm)
Engine	4 – cylinder overhead valve, gasoline, 1997 cc; 4-cylinder 2495cc gasoline; 2.25 litre or 2.50 litre Land Rover Diesel
Transmission	Manual; 4-speed
Fuel capacity	NA
Wheelbase	86 inches (2184mm) to 110 inches (2794mm)
Track	NA
Tire size	NA
Maximum speed (road)	55 mph (90 km/h)
Maximum range	390 miles (628 km)
Maximum gradient	NA
Slope	NA
Ground clearance	NA
Fording depth	NA

A Land Rover Defender maintained by the Defence Forces for ceremonial purposes, photographed in July 2009. Of particular note is the fact that this vehicle is a left-hand drive vehicle, quite in contrast to the normal right-hand drive that is standard on almost all Irish vehicles. (Sean O'Sullivan)

The Land Rover mounted FACE system and operator; note the 4th Field Artillery Regiment badge. (Sean O'Sullivan)

A Land Rover 109 Series III fitted with the FACE fire control system. (Sean O'Sulivan)

Leyland DAF T244

The Leyland DAF 4-tonne 4 x 4 truck was based on a design known as the T244, and is a conventional design with a forward control all-steel cab. The forward tilting cab design was based on that of the Leyland Roadrunner commercial light truck and is configured to allow the vehicle to be driven directly into a C-130 Hercules transport aircraft. The axles for the T244 were specially developed at the Leyland Albion plant in Glasgow. In June 1989 the Leyland DAF was selected by the British Army after testing against competing Bedford and Volvo entries. The Leyland-DAF truck connection is a tale of company mergers, insolvencies, and buyouts. Suffice it to say that Leyland, having run the gamut from British Leyland to Rover Group, to a merger with the Netherlands firm of DAF in 1987, once again became Leyland Trucks in 1993 (and since 1998 has been a subsidiary of the U.S. PACCAR company). Between 1993 and 1998 the Irish Army acquired 55 Leyland 4 x 4 trucks as cab and chassis only, with the truck bodies themselves being built by Cahill Truck Bodies, Graiguenamanagh, County Kilkenny. The trucks were delivered with right-hand drive. The Leylands have spacious cabs and reportedly handle very well. Although the Leylands are referred to as troop carrying vehicles in the Irish Army (with a carrying capacity of 20 troops), they are also used for cargo hauling and as artillery prime movers. The Leylands that have been allotted to artillery units are used as the prime mover for the 105mm Light Guns and the 40mm Bofors anti-aircraft guns. The Leyland is fitted with a dropped hitch that is approximately 26 inches above ground, plus or minus two inches.

A Leyland DAF TCV towing a 25-pdr. The Leylands were bought as cab and chassis, and the bodies were manufactured and fitted by Cahill Bodies, Co. Kilkenny. The 25-pdr is fitted with a canvas muzzle cover to provide some protection against the Irish weather, and has been "dressed" with red paint on its hubs, wheel nuts, and towing shackle. (Sean O'Sullivan)

A 7th FAR convoy of Leylands towing 25-pdrs down a rather narrow Irish country lane in 2006. The 25-pdr has its muzzle and breech covered for protection during transit. (Colin Stone)

An Irish Army Leyland DAF. The Leyland is used extensively as a gun tower, including the 105mm L118 and, prior to their phase out, the 25-pdr. (Sean O'Sullivan)

A DAF 4x4 with one side of the canvas tilt raised, set up as a command post for 62 Reserve Artillery Regiment at the Glen of Imaal on 21 July 2009. The Leyland is primarily used as a troop carrying vehicle and is not organically assigned exclusively as a gun tower in the Irish Army. The CP Truck is normally a stores truck that is a spare, and has been pressed into service as a command post because it provides a fair degree of cover from the weather; its height also offers better sight of the battery and makes control easier. (Colin Stone)

Designation	Leyland DAF T244
Manufacturer	Leyland Trucks, Preston, Lancashire, England
Dates in Irish service	1993 – present
Number in Irish service	55
Weight	23,810 lbs (10,800 kg)
Effective load	9260 lbs (4200 kg)
Length	262 inches (6650mm)
Width	99 inches (2490mm)
Height	135 inches (3430mm)
Engine	Leyland 300 series 6-cylinder turbocharged Diesel, 5.9 liters, 145 HP @ 2600 rpm
Transmission	Turner 350 5-speed syncromesh; 2-speed transfer box
Fuel capacity	29.7 Imperial gallons (35.7 US gallons; 135 liters)
Wheelbase	156 inches (3950mm)
Track	82.68 inches (2100mm)
Tire size	12.00 R20
Maximum speed (road)	55 mph (90 km/h)
Maximum range	310 miles (500 km)
Maximum gradient	60%
Slope	30%
Ground clearance	12.6 inches (320mm)
Fording depth	30 inches (750mm)

A pair of Leylands towing 105mm L 118 light guns, College Green, Dublin, on 16 April 2006. The guns are from the 2nd Field Artillery Regiment's gun battery, stationed at McKee Barracks in Dublin. (Sean O'Sullivan)

A Leyland towing a 105mm light gun, Curragh Camp, in April 2008.The L118 gun has protective coverings on its muzzle and breech mechanism, and both it and the Leyland have highly visible reflectors and reflective strips added for safety purposes. (Sean O'Sullivan)

A Leyland towing a 25-pdr, possibly from 31 Reserve Artillery Regiment, during an exercise in 2005. Muddy field conditions equate to increased post-exercise maintenance and cleanup requirements for both vehicles as well as guns. (Bryan Ely)

A rather muddy DAF with a 25-pdr under tow, taken at Curragh Camp on 27 June 2009. The DAF itself had previously been stuck in mud leaving the gun lines before re-hitching the gun, which accounts for the lack of mud on the tires of the 25-pdr, except in the cleats. (Colin Stone)

MAN 14.240 FAEG

The MAN 14.20 FAEG is a very capable 5-ton 4 x 4 military truck; the first of a fleet of 36 vehicles entered Irish service in 1979, with deliveries culminating in 1989. MAN developed the Category III vehicles based on the Category I vehicles previously developed for the German Army. The Category III vehicles use the same suspension and drive train technology as the Category I vehicles, with a torsionally stiff chassis, coil springs, and axles with planetary gear hub reduction, but use MAN engines instead of Klockner Humboldt Deutz (KHD) engines and commercial tilt cabs. The chassis consists of hollow section longitudinal members welded with tubular cross members, and the rear cargo area has a steel subframe, aluminum drop sides, and a wood floor. Paradoxically, the chief complaint regarding the vehicles by members of the Irish military, who were very impressed by its capabilities, was that not enough of them had been purchased. The MAN 14.240 was used only occasionally to tow artillery pieces, as despite its excellent characteristics, the vehicle's configuration was not ideally suited to towing artillery. The body is very high off the ground, and in

A MAN 14.240 FAEG 5-ton truck, in service with an infantry unit. This type of highly capable truck was used infrequently to tow artillery pieces, as its configuration made it very inconvenient to hitch and unhitch the gun. (Colin Stone)

A 4th S&T (now 4th LSB) MAN 4 x 4 towing a 25-pdr, providing a clear view of the height of the body above ground; the location of the tow hook under the body's overhang made it difficult and sometimes dangerous for crews to hitch the gun. Due to safety concerns, the practice of using the MAN as a gun tower was soon discontinued. The sign on the left rear mudguard cautions against passing on the inside if truck is turning left. (Colin Stone)

A MAN 14.240 FAEG cargo truck with an Irish built OKM 10-ton trailer. The MAN was a very fortuitous purchase by the Irish Defence Forces, providing solid, reliable service in a number of roles. (Colin Stone)

the words of one artilleryman, "the gun crews needed parachutes to exit the back". The tow hook was under the body, which obliged the gun crew to crouch low under the body to hitch the gun to the truck, resulting in several back injuries, added to injuries that some crew sustained from jumping off the back of the body itself. This negative experience led to reverting to other vehicles as gun towers.

Designation	MAN 14.20 FAEG
Manufacturer	Maschinenfabrik Augsburg-Nürnberg (MAN) Ag, Munich
Dates in Irish service	1979 – present
Number in Irish service	36
Weight	18,409 lbs (8350 kg)
Effective load	13,228 lbs (6000 kg)
Length	315 inches (8000mm)
Width	98.4 inches (2500mm)
Height	120.3 inches (3056mm) to top of cab
Engine	MAN D2566 MFG non-turbocharged 6-cyliner in-line 11.41 litre Diesel 240 bhp
Transmission	ZF S6-90
Fuel capacity	68.2 Imperial gallons (81.9 U.S. gallons310 liters)
Wheelbase	177 inches (4500mm)
Track	NA
Tire size	1400-20 (Semperit Gigant Steel)
Maximum speed (road)	55 mph (90 km/h)
Maximum range	NA
Maximum gradient	NA
Slope	NA
Ground clearance	NA
Fording depth	NA

MAN 20:280 DFAEG

The MAN 20:280 is a robust 6 x 6 military truck offered in many variants. The series debuted in the German armed forces in the late 1970s, and like much other German equipment of the period, bore the lessons learned during World War II and was designed to be able to operate under extreme road and weather conditions in Russia. The Irish Army acquired one truck of this type in 1986, mounting the Giraffe 40 radar, to support the Army's 40mm Bofors L/70 anti-aircraft guns and RBS 70 air defence missile systems. It was the only Giraffe system available in Ireland for more than twenty years, until 2008 when a number of improved Giraffe 50 systems mounted on the Swedish Bv 206 were delivered. As of 2012 the MAN/Giraffe system was still in service alongside the Bv 206s.

Designation	MAN 20:280 DFAEG
Manufacturer	Maschinenfabrik Augsburg-Nürnberg (MAN) Ag, Munich
Dates in Irish service	1986 – present
Number in Irish service	1
Weight	22,708 lbs (10,300 kg) empty; 44,754 lbs (20,300 kg) laden
Effective load	22,046 lbs (10,000 kg)
Length	NA
Width	98 inches (2490mm)
Height	118.5 inches (3010mm) to top of cab
Engine	MAN D2566 MTFG turbocharged 6-cyliner in-line 11.41 litre Diesel 280 bhp
Transmission	6 forward, 1 reverse with MAN G801 transfer box with differential lock
Fuel capacity	68.2 Imperial gallons (81.9 U.S. gallons310 liters)
Wheelbase	157.5 inches (4000mm) 1st to 2nd axle; 55 inches (1400mm) second to third axle
Track	NA
Tire size	14.00x20
Maximum speed (road)	55 mph (90 km/h)
Maximum range	NA
Maximum gradient	NA
Slope	NA
Ground clearance	NA
Fording depth	39.4 inches (1000mm)

The Irish Army's solitary 20:280 6 x 6 MAN truck with the Giraffe G40 radar. The radar is in its folded position, and similarly, the stabilizing outriggers are folded in the upright position for travel. This system has been in service with the Artillery Corps since 1986. (National Transport Museum, via Colin Stone)

The Irish Army's MAN 20.280 DFAEG 6 x 6 with its Giraffe radar fully erected. The MAN 20.280 is essentially a larger version of the 5-ton MAN 14.240 FAEG general service truck also used in the Irish Army. (Colin Stone)

Mercedes-Benz 1117A

The Mercedes 1117A was designed as a 4 x 4 cargo carrier, troop transporter, and for towing light artillery pieces. It is a cab-over configuration, with many access panels to the engine and transmission for routine maintenance, and with a tilting cab for more involved maintenance. The Irish Army took delivery of sixty of these vehicles between 2000 and 2002. The trucks were purchased as complete units, i.e., they were delivered with German-built cargo bodies already installed. To comply with Irish standards, the vehicles are fitted with right-hand drive. The 1117A is used as a gun-towing vehicle for the Bofors EL 70 guns, as well as a troop carrier.

A Mercedes 1117A troop carrier photographed in October 2007. Second only to British vehicles, the Irish Army has shown a strong preference for German cargo trucks since the 1960s. (Sean O'Sullivan)

The iconic Mercedes three-pointed star might be considered the only concession to the Spartan appearance of the Mercedes 1117A vehicle. The 1117A was designed for maximum functionality and is well suited to its military duties. (Sean O'Sullivan)

Designation	Mercedes 1117A
Manufacturer	Mercedes-Benz
Dates in Irish service	2000 – present
Number in Irish service	60
Weight	25,346 lbs (11,500 kg)
Effective load	NA
Length	275 inches (6980mm)
Width	98 inches (2480mm)
Height	109 inches (2768mm)
Engine	Mercedes-Benz OM 366A turbocharged 6-cyliner Diesel. 170 HP @ 2600 rpm
Transmission	6-speed G3/55-6/8.5 with VGW 500-3W transfer case
Fuel capacity	44 Imperial gallons (52.8 US gallons, 200 liters)
Wheelbase	143 inches (3620mm)
Track	NA
Tire size	365/85 R20
Maximum speed (road)	61 mph (98 km/h)
Maximum range	NA
Maximum gradient	40% on road; 80% off-road
Slope	NA
Ground clearance	NA
Fording depth	NA

A Mercedes 1117A demonstrating its ability to climb a steep grade in a cross-country environment. Its off-road climbing ability is quite impressive, both with and without a trailed load. (Paul Murphy)

A Mercedes 1117A with a Bofors L70 under tow, taken at the Curragh workshops in July 2006. The Mercedes was often used to tow the Bofors L/70 guns and the associated Flycatcher radar trailers. (Colin Stone)

Mercedes-Benz LA 911B

The Mercedes LA 911B was a 4 x 4 militarized commercial truck, produced as the 911 beginning in 1963. It remained in production for many years and was in widespread military use throughout the world. Seven examples of the LA 911B entered Irish service in 1971 and served until 1990; all seven were later scrapped. While in service, they were parceled out as towing vehicles to the 6th Field Artillery Regiment, Kildare for towing the 25-pdr gun/howitzer, and the 1st Air Defence Regiment for towing the 40mm Bofors guns.

A battery of 25-pdrs from the 6th Field Artillery Regiment training in the Glen of Imaal in 1985. Behind the guns is a Mercedes LA 911B, used as a gun tower for the 25-pdrs. (62 Reserve Artillery Regiment, via Capt Ian Stewart)

A pair of LA 911B trucks participating in a 6th FAR parade. (Colin Stone)

Four Mercedes 911B trucks associated with the 6th Field Artillery Regiment as prime movers for one of the regiment's 25-pdr batteries during a 1985 exercise. (62 Reserve Artillery Regiment, via Capt Ian Stewart)

Designation	Mercedes LA 911B
Manufacturer	Mercedes-Benz
Dates in Irish service	1971 -1990
Number in Irish service	7
Weight	12,125 lbs (5,500 kg) empty; 22,046 lbs (10,000 kg) loaded
Effective load	9,921 lbs (4,500 kg)
Length	257 inches (6590mm)
Width	97 inches (2464mm)
Height	129 inches (3300mm)
Engine	Mercedes OM 352, 6-cylinder Diesel, 5675cc, 110 HP @ 2900 rpm
Transmission	MB G3/50-5/8.5. 5-speed, with 2-speed transfer case
Fuel capacity	29.7 Imperial gallons (35.7 US gallons; 135 liters)
Wheelbase	142 inches (3600mm)
Track	77.6 inches (1970mm)
Tire size	11.00 – 20
Maximum speed (road)	52.2 mph (84 km/h)
Maximum range	155 miles (250 km)
Maximum gradient	68%
Slope	NA
Ground clearance	NA
Fording depth	35.4 inches (900mm)
Special notes	Front-mounted PTO drum winch

Mistubishi Pajero LWB

The tender for a 4 x 4 vehicle to replace the Nissan Patrol series of vehicles was issued in 2008. Vehicles were offered for consideration by Kia, Land Rover, Nissan, Mitsubishi, and Toyota. Extensive testing was carried out on the vehicles to determine their suitability for military use. The winner of the competition, which considered factors such as reliability, cost, and serviceability was the Mitsubishi Pajero LWB (Long Wheelbase), and in May 2009 a contract valued at 10 million Euros was awarded. There are varying interpretations of the Pajero designation, ranging from "one who carries or sells straw," to a type of Argentinian pampas cat, to some more earthy colloquial expressions. The first 80 Pajeros were delivered between May and June 2009, with a total possible purchase of up to 320 over the following four years. The Irish Army Pajeros will vary only slightly from the standard civilian version of the vehicle; chrome or other shiny ornamentation will be lacking, and the wheels are simple pressed steel wheels rather than alloy wheels. Standard features include the Mitsubishi Super Select four-wheel drive system, traction control, air conditioning, cruise control, and electric windows. A significant amount of modification was required to integrate the military communications systems, and another specific addition is a NATO-specification tow hitch which will enable the Pajero to tow the 120mm RUAG mortar, among other loads. The Pajeros will be used primarily for patrol and communications purposes, cash in transit and prisoner escorts, driver training, and on overseas peace support operations.

One of the first lot of Pajeros delivered to the Irish Defence Forces, shown inside the Combined Vehicle Base Workshops in May, 2009, in pristine condition. (Ralph Riccio)

A Mitsubishi Pajero towing an Irish-built RBS 70 trailer, May 2011. The Pajero is also used as a gun tower for the RUAG M87 120mm mortar. (Colin Stone)

Four Pajeros of the first batch to be delivered to the Irish Defence Forces, shown in front of the Mitsubishi Motors Ireland showroom in May, 2009, prior to actual delivery. Supports for the radio antenna have been factory installed. The Pajeros are factory painted in Matt Olive Drab, which is used exclusively by the Defence Forces. (Mitsubishi Motors Ireland)

Designation	Mistubishi Pajero LWB
Manufacturer	Mitsubishi Motors, Tokyo, Japan
Dates in Irish service	2009- present
Number in Irish service	80; possible total of 320
Weight	6680 lbs (3030 kg) GVW
Effective load	NA
Length	173 inches (4385mm)
Width	74 inches (1875mm)
Height	73 inches (1850mm); 74 inches (1880mm) with roof rail
Engine	3.2 litre DI-D 16-valve Common Rail Intercooled Diesel, 200 HP
Transmission	5-speed manual
Fuel capacity	19.4 Imperial gallons (23.25 US gallons; 88 litres)
Wheelbase	100 inches (2545 mm)
Track	62 inches (1570mm)
Tire size	265-65 R17
Maximum speed (road)	110 mph (177 km/h)
Maximum range	NA
Maximum gradient	NA
Slope	NA
Ground clearance	NA
Fording depth	NA
Special notes	Towing capacity is 7275 lbs (3300) kg with a braked trailer, 1653 lbs (750 kg) with an unbraked trailer

A Pajero on an unimproved track. The vehicle is a no-frills version of the standard civilian model of the Mitsubishi Pajero. (Department of Defence).

This Pajero seems to be well up to the challenge of negotiating a muddy stream. Presumably, this activity represents vehicle testing rather than normal operational practice. (Department of Defence)

Morris-Commercial CDSW 6 x 4 artillery tractor

The Morris-Commercial CDSW was the last 6-wheeled truck built by Morris. It was first manufactured in 1935 and had been developed to tow the 18-pdr and 4.5 inch howitzer, but later was pressed into service as a prime mover for the 25-pdr gun/howitzer. It was superseded in British service by the Morris Quad 4 x 4 FAT. The designation CDSW was a "shorthand" denoting C as the model number, D for dual-axle, S for six-cylinder, and W for fitted with winch. In 1938 and 1939 the Irish Army procured a total of nineteen of the CDSWs and fitted them with bodies built at Clancy Barracks, based on the standard British design. The CDSWs in Irish service were used almost exclusively to tow the twelve 17-pdr anti-tank guns, although one was reserved as a gun carriage for military funerals. The CDSWs were supplemented as gun towers by the Morris Quad Mk IIs, and were disposed of in 1965 and 1969.

CDSW ZC 5830 towing a 17-pdr anti-tank gun, sometime in the 1950s. The CDSW was somewhat of a jack-of-all trades as a gun tower, used to tow virtually every type of gun in the Irish inventory at the time, including field artillery and anti-tank artillery. (Military Archives)

Two Morris CDSW tractors towing 4.5 inch howitzers and their limbers. Although the registration numbers ZC 5830 and ZC 5831 of the CDSWs are consecutive, it is interesting to note the different letter and number styles of the two plates. (MA, Hanley Collection G120)

A Morris-Commercial CDSW 6x4 truck towing a 4.5 inch howitzer in a parade in Dublin in the 1950s. (Military Archives, via Denis McCarthy)

Morris-Commercial CDSW 6x4 trucks towing 18-pdrs during a parade held in Dublin. The trucks behind the CDSWs are 1940 pattern Ford V-8s. (Military Archives via Denis McCarthy)

A pair of CDSWs on parade with artillery under tow in the 1940s. The CDSWs are followed by Ford 1940 pattern trucks, which in turn are followed by Ford 1939 pattern trucks. (Lt Col Edmond Fogarty)

Morris-Commercials ZD 297 and ZD 288 towing 25-pdrs, passing in front of the GPO in Dublin during the Republic Day Parade in 1948. They are followed by what appears to be a Ford "Bug Chaser", which in turn is followed by another group of CDSWs, and, further behind, by Morris Quads. (Shay Stone via Colin Stone)

An 18-pdr being readied for towing by Morris-Commercial CDSW ZD 288 at the Leitrim gun position in 1963. (Shay Stone via Colin Stone)

CDSW ZD 288 also was used for more somber purposes than in its accustomed role as a gun tower. Here it is being used as a funeral carriage or an officer's funeral, date unknown. (Shay Stone viaColin Stone)

Designation	Morris-Commercial CDSW 6 x 4 artillery tractor
Manufacturer	Morris-Commercial, Birmingham, England
Dates in Irish service	1938 – 1969
Number in Irish service	19
Weight	8148 pounds (3696 kg)
Effective load	NA
Length	198 inches (5029mm)
Width	80 inches (2032mm)
Height	73 inches (1850mm); 74 inches (1880mm) with roof rail
Engine	Morris 6-cylinder side-valve, petrol, 3485cc, 60 HP
Transmission	5-speed manual
Fuel capacity	22 Imperial gallons (26.5 US gallons; 100 liters)
Wheelbase	115.5 inches (2934mm) front axle to center of rear bogie; bogie centers 40 inches (1016mm)
Track	NA
Tire size	9.00 x 16
Maximum speed (road)	NA
Maximum range	NA
Maximum gradient	NA
Slope	NA
Ground clearance	10.5 inches (267mm)
Fording depth	NA
Special notes	Fitted with 2-inch circular winch with 120 feet of steel cable; winch could pull from front or from rear

The immaculately restored ex-Irish CDSW at Coolmoney Camp in July 2009, participating in the pageantry surrounding the final 25-pdr shoot by the Irish Artillery Corps. The entire combination of CDSW, limber for the 25-pdr, and the 25-pdr field piece itself has been assembled, marking the first time in years that this combination had been seen together in the Glen. The gun tower is shown here with its British registration plate 225 UXW; other photographs of the vehicle on the same occasion show it with a plate with its original Irish registration number, ZD 288. (Jonathan Catton, via Colin Stone)

Morris Commercial CDSW artillery tractor ZC 5835 as restored by The Garrison, shown on the occasion of the final 25-pdr shoot at Glen of Imaal on 30 July 2009. (Colin Stone)

Morris-Commercial C8 4 x 4 FAT

The British Morris-Commercial C8 FAT (Field Artillery Tractor), commonly referred to as the Morris Quad, was a somewhat ugly looking 4 x 4 artillery tractor commonly used to tow the 25-pdr gun/howitzer and the 17-pdr anti-tank gun. Its unique configuration with its sloping back has been described as "beetle-shaped." The original 1937 design for the tractor was by Guy on a Morris-Commercial C8 4 x 4 chassis, and from 1938 throughout the war years, the FAT was built in numerous versions by several manufacturers including Guy and Karrier. In 1942 the Irish Army took delivery of 35 Quad Mk II artillery tractors. The Quads were used to tow various types of field artillery pieces, as well as being temporarily used to tow anti-aircraft guns. The Quads served until the 1960s, when they were replaced as prime movers by the more conventional Bedford RL trucks.

The vehicles in this photograph – at least two Morris CDSWs and one 4x4 Quad – are almost totally obscured by the gunners from 19th Battery, 7th FAR. The picture dates from the 1961 Camp. Notice the Bush radio, the youthful faces, and the overcoats in the middle of the summer. (Colin Stone)

Morris-Commercial 4x4 Quad field artillery tractors towing 25-pdr guns at an Easter Parade in Dublin. Despite their ungainly appearance, the C8s were well regarded in the role they were designed for, specifically as artillery tractors. (Military Archives, via Denis McCarthy)

Three Morris-Commercial C8 Quad artillery tractors, date unknown. The original caption to this photograph identifies the individuals as Cpl J. Miley, Gnr McLoughlin, and Cpl Kelly. (62 Reserve Artillery Regiment, via Capt Ian Stewart)

Designation	Morris-Commercial C8 4 x 4 FAT
Manufacturer	Morris-Commercial, Birmingham, England
Dates in Irish service	1942 – 1960s
Number in Irish service	35
Weight	7500 lbs (3402 kg)
Effective load	NA
Length	177 inches (4490mm)
Width	87 inches (2210mm)
Height	89 inches (2260mm)
Engine	Morris EH, 4-cylinder gasoline, 3520 cc 70 HP @ 3000 rpm
Transmission	5-speed
Fuel capacity	30 Imperial gallons (36 US gallons; 136 liters)
Wheelbase	99 inches (2525mm)
Track	NA
Tire size	10.50 x 20 (early versions) 10.50 x 16 (later versions)
Maximum speed (road)	50 mph (80 km/h)
Maximum range	160 miles (257 km)
Maximum gradient	NA
Slope	NA
Ground clearance	NA
Fording depth	NA
Special notes	4-ton winch standard

C8 Quad ZD 3177 at the National Transport Museum, Howth. The Quad, along with another Morris-Commercial vehicle, the CDSW, provided a significant portion of the towing capacity for the guns in the Artillery Corps during The Emergency. (National Transport Museum via Colin Stone)

Nissan Patrol 260 Series RD28

The Nissan Patrol is a four-wheel drive SUV that was first introduced in 1951, manufactured for military use. The original vehicle has undergone continuous upgrading and modernization, retaining the Patrol badge much as have similar vehicles such as the Jeep and the Land Rover. As of 2010, the Irish Army used various versions of the Nissan Patrol as its principal utility vehicle. The 260 series RD28 was first delivered in 1989 as a replacement for the Land Rovers then in service; the RD28 is powered by a Diesel engine. All were modified for military use, and many were fitted with radios. Among other duties, they are used for towing support weapons. The newer GR TD-6 has been in service since 2000, although some of the 260 series still remain in service; the 260 series had far better off-road capabilities than do the newer GR series. One drawback to the 260 series

A Nissan Patrol 260 taken in Curragh Camp, October 2005. Although being replaced by newer vehicles, the 260 series, built specifically for military use, had better off-road capabilities than some of the newer replacement vehicles. The Nissan is parked next to one of the Irish Army's two SISU 6 x 6 armoured personnel carriers. (Sean O'Sullivan)

A somewhat tired-looking Nissan Patrol 260 in October 2005. The bonnet has been repainted in a markedly different shade of green than the rest of the vehicle. (Sean O'Sullivan)

Mullingar, Artillery Day 1998. A Nissan Patrol 260 RD 28 towing a trailer used to transport the RBS 70 air defense missile system. (Colin Stone)

Nissan Patrol 260 series RD 28 with a RUAG M87 mortar under tow. Although the RUAG is a rather large mortar, it presents a very compact appearance on its trailer. The layout of the trailer reflects its Swiss heritage of precision, with tools and parts for all occasions arranged in an exact position on the carriage. (Colin Stone)

was that the spare wheel was carried on the underside of the vehicle and was prone to being torn off while crossing obstacles. As a consequence, the spare wheel often was dismounted and stowed in the back of the vehicle, sharing space with the radio. Although the production line for the 260 had been shut down, Nissan reopened the line to fulfill additional orders by the Irish Army; the later acquisitions were not fitted with a 24-volt electrical system for the ANPRC 46 radio, leading to use of a SINCGARS manpack radio system with a jury-rigged antenna.

Designation	Nissan Patrol 260 Series RD28
Manufacturer	Nissan Motor Company Ltd., Tokyo, Japan
Dates in Irish service	1989 – present
Number in Irish service	NA
Weight (unloaded)	4453 lbs (2020 kg)
Effective load	1720 lbs (780 kg)
Length	188 inches (4770mm)
Width	68 inches (1734mm)
Height	80.5 inches (2045mm)
Engine	Nissan RD28 6-cylinder in-line Diesel, 2826cc 130 HP
Transmission	FS5R30A 5-speed manual, with two-speed transfer box
Fuel capacity	18 Imperial gallons (21.7 US gallons; 82 liters)
Wheelbase	118 inches (3000mm)
Track	NA
Tire size	205 SR 16
Maximum speed (road)	90 mph (145 km/h)
Maximum range	500+ miles (805+ km)
Maximum gradient	NA
Slope	NA
Ground clearance	7 inches (175mm)
Fording depth	NA
Special notes	4-ton winch on very few vehicles

Nissan Patrol GR TD-6/GR 3.0 Di

In 2000, about 100 Nissan Patrol GR TD-6 were purchased and began to replace the older Nissan Patrol 260 series. The TD-6 did not prove to be satisfactory, and soon thereafter, in 2001, the GR 3.0 Di, with a more powerful engine and more torque, began to be delivered. Subsequent deliveries were made in 2003, 2004, 2005, and 2008, as well as possibly in 2006 and 2007. Although the vehicle was easy to drive on paved roads, was fast, was well suited for the Aid To Civilian Power missions and was popular overseas, its low ground clearance and unsuitable tires contributed to poor off-road performance compared to the earlier Nissan Patrol 260. The GR was configured to carry five troops with the radio operator sitting on a seat in the rear compartment, but legroom was virtually nonexistent, so signalers avoided using the seat. Administratively, introduction of the GR 3.0 caused problems as the 3.0 litre Diesel engine required drivers to undergo conversion courses as the engine was above the legal rating allowed for General Service drivers (the previous 2.8 litre Diesel was within the rating). In 2009, Nissan was one of the competitors for a new tender of vehicles, but the competition was won by the Mitsubishi Pajero 4 x 4, which will supplant and eventually replace the Nissans in service.

A 6-cylinder 2.8 litre Nissan Patrol GR TD. The GR TD was an interim vehicle between the older Patrol 260 and the newer Patrol GR. It was not as reliable as the Patrol 260, and its intercooled diesel engine developing 130 HP was underpowered compared to the GR. (Sean O'Sullivan)

A 4-cylinder 3.0 litre Nissan Patrol GR, 14 February, 2008. The GR was widely used in the Irish Army, being first introduced in 2001; in 2009 the Mitsubishi Pajero was chosen to supplant the GR. A number of GRs were fitted with left-hand drive for service abroad; the GR pictured is one of the LHD vehicles. The trailer hitch seen on the rear is used to tow the 120mm mortar. (Sean O'Sullivan)

A 120mm RUAG in the travel configuration, being towed by a Nissan Patrol GR 3.0 Di, at the Glen of Imaal on 12 September, 2009. The mortar itself is under a protective tarpaulin, but the massive base plate is evident. The relatively recent adoption of the RUAG probably accounts for the apparent lack of wear on the base plate. (Roy Kinsella)

Designation	Nissan Patrol GR TD-6/GR 3.0 Di
Manufacturer	Nissan Iberia, Barcelona, Spain
Dates in Irish service	2001 – present
Number in Irish service	NA
Weight (unloaded)	5291 lbs (2400 kg)
Effective load	1500 lbs (680 kg)
Length	197 inches (5010mm)
Width	76 inches (1930mm)
Height	80.5 inches (2045mm)
Engine	Nissan ZD30 DDTi 4-cylinder inter-cooled turbo-Diesel, 156 HP
Transmission	5-speed manual with two-speed transfer box
Fuel capacity	20.9 Imperial gallons (25 US gallons; 95 liters)
Wheelbase	117 inches (2970mm)
Track	63 inches (1605mm) front; 64 inches (1625mm) rear
Tire size	265/70 R16 112S
Maximum speed (road)	99 mph (159 km/h)
Maximum range	548 miles (882 km)
Maximum gradient	NA
Slope	NA
Ground clearance	8.5 inches (215mm)
Fording depth	NA

A Nissan Patrol GR with a 120mm RUAG M87 mortar under tow, June 2011. The RUAG is a highly mobile unit, stowed on its specially designed trailer. (Colin Stone)

Scania R420 CB 6 x 6

In September 2006 the Irish Defence Forces announced that it had awarded a contract for 36 Scania R420 CB trucks to be used as troop carriers. The majority of the vehicles were delivered in late 2006 and early 2007, painted in RAL 6014 green. The R420 layout is a cab-over-engine configuration, fitted with CR 19 sleeper cabs. The bodies were built by EMPL in Germany and Austria. The trucks can be quickly converted to carry a standard 20' ISO container if necessary. The vehicles are fitted with right-hand drive for Irish service, and incorporate EGR (Exhaust Gas Recirculation) technology. Although the R420s are not dedicated exclusively to tow artillery, they are capable of doing so as required.

A pair of Irish Army Scania R 420 6 x 6 trucks at the Glen of Imaal in July 2009. Although the truck chassis is built in Sweden, the bodies are furnished by a manufacturer based in Germany and Austria. (Roy Kinsella)

As of early 2012 the Scanias were the most modern general service truck in the Irish Army inventory. The Scania is the first 6 x 6 cargo truck to be adopted by the Irish Army, a departure from the 4 x 4 trucks that had been the norm prior to this purchase. A significant number have been committed to support Irish peacekeeping operations abroad. (Roy Kinsella)

A Scania 6x6 at the Curragh in December 2009. Note the "windows" in the tilt covering. (Colin Stone)

Designation	Scania R420 CB 6 x 6
Manufacturer	Scania, Södertälje, Sweden
Dates in Irish service	2006 – present
Number in Irish service	36
Weight (unloaded)	28,660 lbs (13,000 kg)
Effective load	68,343 lbs (31,000 kg)
Length	354 inches (9000mm)
Width	102 inches (2600mm)
Height	144 inches (3662mm)
Engine	Scania Euro 4 6-cylinder inline Diesel (with Scania EGR), 420 @1900 rpm
Transmission	HD 4500 Allison 6-speed automatic
Fuel capacity	88 Imperial gallons (106 US gallons; 400 liters)
Wheelbase	169 inches (4300mm) between 1st and 2nd axle; 53 inches (1355mm) between 2nd and 3rd axle
Track	NA
Tire size	14.00 R20, Michelin XZL
Maximum speed (road)	No limiter
Maximum range	NA
Maximum gradient	NA
Slope	NA
Ground clearance	NA
Fording depth	NA

A Scania R 420 in December 2007. The Scanias delivered for Irish service are fitted with right-hand drive, and incorporate many technologically advanced power train features. The 6 x 6 layout enhances off-road capabilities. (Sean O'Sullivan)

Scania R 420 CB trucks with the Irish contingent in Chad in 2009. This lineup of fifteen trucks represents nearly half of the total number of Scanias in the Irish inventory. The fourth truck from the left appears to be the only one missing its protective mesh grille in front of the radiator. (Irish Defence Forces)

Universal Carrier MK I/Mk II

The Universal Carrier was a light tracked vehicle developed in the 1930s in Britain as a machinegun carrier. It was commonly referred to as the Bren Carrier in reference to the armament with which it was fitted. The hull of the vehicle was a simple steel box with the engine compartment located in the center. The driver and gunner sat in a front compartment, with the radiator grill mounted in a bulkhead between them. Behind the crew were two rectangular compartments, one on each side of the engine, and were used to carry supplies or personnel. In 1940, the Irish Army took delivery of 26 Universal Carriers, Mk I, all of which were assigned to the Cavalry Corps' newly formed Carrier Squadron. In order to fully equip the infantry battalions with carriers, a further two hundred carriers were purchased in 1942, among them a number of mortar carriers. Most battalions received nine Universal Carriers, four of which were for the mortar platoon. The mortar carriers were used to carry the British 3-inch mortar or the French Brandt 81mm mortar. The mortar tube was carried mounted horizontally, secured in brackets on the rear plate, and the bipod legs and base plate were carried on a platform underneath the mortar tube, with the base plate to the right. Ammunition was stowed vertically in the troop compartment. Around 1949, some of the tracked carriers were used to tow the 6-pdr anti-tank guns.

Rear view of a 3 inch mortar carrier showing stowage arrangement for the mortar tube, bipod, and base plate. Also visible is the ammunition stowage arrangement in what originally served as the troop compartment. Vertical storage of the ammunition allowed 66 rounds to be carried. (Defence Force Regulations, Manual of the Mortar [81m.m. and 3-inch])

Stowage arrangements for an 81mm mortar on the rear of a Universal Carrier. The tube, bipod, and base plate are stowed much as they were for the 3 inch mortar on the Universal Carrier. Ammunition is stored vertically in the troop compartment, providing easy access. (Defence Force Regulations, Manual of the Mortar [81m.m. and 3-inch])

A Mk I Universal Carrier towing a 6-pdr anti-tank gun. Interestingly, the gun carries branches as camouflage, suggesting that it was in a wooded position, possibly in the trees to the right of the photograph. (Military Archives, via Denis McCarthy)

Designation	Universal Carrier MK I/Mk II
Manufacturer	Various (Vickers-Armstrong, Bedford, Wolseley, Thornycroft, Morris, Ford, Aveling)
Dates in Irish service	1940-1944 (Irish Defence Forces)
Number in Irish service	3226 total (number of mortar carrier version not determined)
Crew	4 (including mortar crew)
Combat weight	7500 lbs (3402 kg) standard carrier; 9630 lbs (4368 kg) mortar carrier
Armament	3-inch Mk II mortar or 81mm Brandt M44 mortar
Ammunition capacity	66 rounds (3-inch)
Length	144 inches (3650mm)
Width	83 inches (2110mm)
Height	62 inches (1570mm)
Engine	Ford V-8, gasoline, 3631cc, 85 HP
Transmission	Manual, 4 forward, 1 reverse
Fuel capacity	20 Imperial gallons (24 US gallons; 91 liters)
Maximum speed (road)	30 mph (48 km/h)
Maximum radius (road)	150 miles (241 km)
Ground clearance	8 inches (203mm)
Trench crossing	NA

Appendices

Appendix 1: Irish Artillery Corps Insignia

1st Field Artillery Regiment (PDF), 1st Southern Brigade

2nd Field Artillery Regiment (PDF), 2nd Eastern Brigade

2nd Field Artillery Regiment (PDF) (variant)

3rd Field Artillery Regiment (FCA), 1st Southern Brigade (blue number/ gun)

3rd Field Artillery Regiment (FCA), 1st Southern Brigade (black number/gun)

4th Field Artillery Regiment (1970-1976)

4th Field Artillery Regiment (PDF), 4th Western Brigade

5th Field Artillery Regiment HQ (FCA), 4th Western Brigade

Cadre, 5th Field Artillery Regiment (FCA), 4th Western Brigade

6th Field Artillery Regiment (FCA)

6th Field Artillery Regiment (FCA),
Defence Forces Training Centre

7th Field Artillery Regiment (FCA),
2nd Eastern Brigade

8th Field Artillery Regiment (FCA),
1st Southern Brigade

8th Field Artillery Regiment (FCA),
1st Southern Brigade (variant)

8th Field Artillery Regiment (FCA),
1st Southern Brigade (variant)

9th Field Artillery Regiment (FCA),
4th Western Brigade

9th Field Artillery Regiment (FCA),
4th Western Brigade (variant)

9th Field Artillery Regiment (FCA),
4th Western Brigade (subdued)

Depot Artillery, Defence Forces Training Centre

Coast Defence Artillery

Coast Defence Artillery (FCA)

2nd Battery, 1st Field Artillery Regiment
(1973-1975)

2nd Battery, 1st Field Artillery Regiment
(1973-1975) (variant)

4th Field Artillery Regiment
(1970-1976)

4th Battery, 5th Field Artillery Regiment
(1968-2005)

7th Battery, 5th Field Artillery Regiment

13th Battery, 5th Field Artillery Regiment

13th Battery, 5th Field Artillery Regiment
(1968-2005)

15th Battery, 4th Field Artillery Regiment
(1966-1970)

16th Heavy Mortar Battery, 3rd Field Artillery Regiment
(1969-1972)

20th Battery, 4th Field Artillery Regiment
(1966-1970)

20th Battery, 9th Field Artillery Regiment

20th Battery, 9th Field Artillery Regiment
(variant)

21st Heavy Mortar Battery,
1st Field Artillery Regiment (1972-1976)

The insignia shown in this appendix are provided courtesy of Flt Sgt James Perkins, Irish Air Corps

Appendix 2: Artillery Service Dates by Type

Designation	Period
Ordnance QF 18-pdr, Mks I though V	1922 – 1974
Stokes 3-inch trench mortar	1922 – (?)
Ordnance QF 4.5 inch howitzer	1925 – 1976
QF 12 pounder 12 cwt coast defence gun	1938- (still in use as salute guns in 2012)
QF 4.7 inch Mk IV naval gun	1938 – 1940s
BL 6 inch Mk VII naval gun	1938 – 1950?
BL 9.2 inch gun Mark X	1938 – 1969?
QF 3.7 inch anti-aircraft gun	1939 – 1946 (?)
QF 40mm Mk III (Bofors L/60) anti-aircraft gun	1939 – 2001
QF 3 inch 20 CWT anti-aircraft gun	1940 – 1946 (?)
Canon de 75, modèle de 1897	1941 – 1968
Ordnance BL 60 pounder gun	1942 – 1946 (?)
Ordnance ML 3-inch mortar (Mk 44?)	1943 – 1981
3.7 inch howitzer	1933 – 1977
Mortier de 60mm Modèle de 1935 (Brandt)	?
Ordnance QF 2-pdr anti-tank gun	1939 – 1959 (?)
Ordnance QF 6-pdr anti-tank gun	1949 – 1962 (?)
Ordnance QF 17-pdr anti-tank gun Mk I	1949 – 1962 (?)
Ordnance QF 25-pdr field gun	1949 -2009
Brandt AM 50 120mm mortar	1953 – 2008
L 118/119 Light Gun	1980 – present
Hotchkiss-Brandt MO-81-61-L 81mm mortar	1981 – present
Bofors 40mm L/70 anti-aircraft gun	2001 – present
Denel Vektor M1 60mm mortar	2003 – present
RUAG 120/81mm mortar	2008- present

Appendix 3: Surviving Guns

A number of vintage guns that served with the Irish Army have been preserved, some in rather unlikely places. The following list, believed to be reasonably complete, does not pretend to be definitive.

Canon de 75, modèle de 1897

One example has been restored and is maintained by the Cavan & Leitrim Railway Museum, Dromod, County Leitrim. Two are displayed at Columb Barracks, Mullingar, and at least two other examples are maintained at Curragh Camp, but are in rather deplorable condition.

A restored French 75 at Dromod, October 2007. This gun, serial number 15004, was built in Tulle in 1917. It was purchased as scrap from the Irish Army and restored as shown by the Cavan & Leitrim Railway. The wood spoke wheels are reproductions, but do not faithfully replicate the originals which have fourteen spokes and a completely different hub configuration. (Philip Bedford)

One of the French 75's formerly in Irish service, at Infantry Square in the Curragh. Sadly, the piece is in rather sorry condition and, aside from being painted in rather dubious colors, there appears to have been no serious attempt at restoration. Close inspection shows that there are serious corrosion problems. The wood spoke wheels have been retrofitted, as during The Emergency the 75s were fitted with modernized carriages and pneumatic tires. The standard French wood spoke wheels normally had fourteen spokes, while the wheels on this gun, the wood of which is also deteriorating, have only twelve spokes, and appear to be possibly wheels from an old 18-pdr or an 18-pdr limber. (Sean Reilly, via Ian Stewart)

A pair of French 75s at the 4th Field Artillery Regiment's Columb Barracks in Mullingar in September 2011. The carriages of both guns have been raised from the ground on jacks to relieve stress on the wheels themselves. (Colin Stone)

Ordnance QF 18-pdr

A number of 18-pdrs have survived, but are all the later versions mounted on modernized chassis. None of the Mk I or Mk II versions have survived. One Mk IV belonging to the Artillery Corps is on display at Curragh Camp, and another totally restored Mk IV is also located at Curragh Camp; the Cavan & Leitrim Railway museum has two 18-pdrs, one manufactured in 1919 that has been restored, and one from 1926 that as of 2009 was awaiting restoration; another 18-pdr, Mk unknown, is on display at the former Ballincollig Artillery Barracks (now converted to a shopping centre). The National Museum of Ireland, Collins Barracks, Dublin, has a "hybrid" 18-pdr gun on display whose pneumatic tires have been replaced by the earlier wooden spoke artillery wheels to render it appropriate for its WW I exhibit, but the gun tube itself is a post WW I Mk IV. Two 18-pdrs on display at Kickham Barracks, Clonmel, were removed in late 2009 and replaced by two 25-pdrs; as of early 2010 they may have been once again relocated to Kickham Barracks.

The splendidly restored Mk IV 18-pdr of the Curragh Museum, 8 May 2011. This excellent restoration highlights the gun's brass parts by brightly burnishing them, although in service these parts were most often painted over to match the rest of the gun and to make the gun less conspicuous. (Colin Stone)

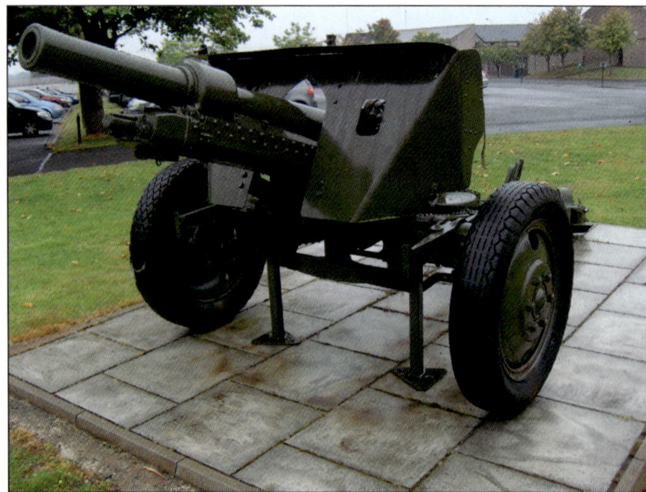

An 18-pdr from the front, Curragh, 9 April 2010. The two legs supporting the axle relieve stress on the tires, each of which is of a different pattern. (Colin Stone)

The Cavan & Leitrim Railway's restored 1919 18-pdr at a military day display. Magnification of this image shows what appears to be part of the Martin Parry company name cast into the right axle. The restoration is rendered all the more impressive by the attention to detail such as the inclusion of leather accoutrement bags. (Philip Bedford)

The 18-pdr on display at Collins Barracks, Dublin. This gun is a hybrid, using a Mk IV gun on a Mk IV carriage retrofitted with wooden artillery wheels from a Mk I or Mk II gun. (Colin Stone)

Ordnance QF 25-pdr

The 25-pdrs were retired from service in July 2009, but six have been retained by the Artillery Corps for ceremonial purposes at McKee Barracks, Dublin. One is on display at Belvedere House, Mullingar, one is at Fort Dunree, one is at Arbour Hill, Dublin, and as of 2012 two from the Glen of Imaal are being repainted at the Curragh.

A 25-pdr in ceremonial livery prepared for the launch of a Presidential ceremony in 2005. This is one of the six 25-pdrs, with possibly two in reserve, similarly maintained for ceremonial occasions; the 25-pdr was phased out of the operational inventory in July 2009. The 25-pdrs retained for ceremonial purposes are recognizable by their chromed muzzle brake as well as by the high visibility appurtenances such as the shovel and white rope. The ceremonial guns are also identifiable by the identifying letters on the gun shield, below the Artillery Corps logo. (Sean Murphy)

A 25-pdr reserved for ceremonial duties photographed on the occasion of the final 25-pdr shoot on 30 July 2009. The guns themselves, as well as the associated issue tools such as the shovel, are maintained in immaculate condition. (Sean Murphy)

A 25-pdr presented to the Irish U.N. Veterans, and located at Arbour Hill, Dublin. The gun is resting on its circular firing platform, but nevertheless is also supported by metal uprights to relieve stress on the tires. The photograph was taken in May 2011, not long after refurbishment of the gun as a memorial. (Colin Stone)

A 25-pdr being refurbished at Curragh Camp in December 2011. Final disposition of this gun when refurbishment is complete is unknown at the time of writing. To the left of the 25-pdr is a 17-pdr also being refurbished. (Colin Stone)

3.7 inch mountain howitzer

One 3.7 inch mountain howitzer has been restored and is maintained by the Artillery Corps at the Defence Forces Training Centre in Curragh Camp. A second 3.7 inch howitzer has been restored and is maintained by the Cavan & Leitrim Railway Museum. Two others are stored at McKee Barracks, Dublin, but are in dire need of restoration.

The Artillery Corps' restored 3.7 inch mountain howitzer in storage at Curragh Camp, April 2006. This is one of about a dozen surviving examples of this type of gun world-wide, at least two more of which are in Ireland. (Bryan Ely)

The 3.7 inch mountain howitzer as restored by the Artillery Corps, in an 8 May 2011 photo. The original carriage of the 3.7 was fitted with wooden spoke wheels, upgraded with pneumatic tires by World War II. It is fitted with the frontal shield, but not with the side shields. (Colin Stone)

Rear view of the Dromod 3.7 inch mountain howitzer. This gun was manufactured in 1918. A great deal of attention was lavished on restoration of this gun by volunteer workers. (Philip Bedford)

A 3.7 inch howitzer stored at McKee Barracks, Dublin, in October 2009. Although restorable, it is in need of serious attention to bring it up to standard suitable for display. (Sean Reilly, via Ian Stewart)

Ordnance QF 4.5 inch howitzer

At least two 4.5 inch howitzers are maintained on display at McKee Barracks, Dublin; a third may also be maintained by the Artillery Corps at the Defence Forces Training Centre in Curragh Camp.

A 4.5 inch howitzer at Cathal Brugha Barracks, 31 August 1998, at a display marking the 50th anniversary of the "End of The Emergency." The upper portion of the shield is raised for the display; the upper shield on the 4.5 inch is normally seen in a lowered position to the front. (Denis McCarthy)

A 4.5 inch howitzer at McKee Barracks, Dublin, in October 2009. Partially visible to the right of the gun is a second 4.5 inch howitzer, similarly restored and painted. (Sean Reilly, via Ian Stewart)

Ordnance QF 2-pdr anti-tank gun

Four 2-pdrs are known to have survived, one at Cathal Brugha Barracks, Dublin, one at the Curragh (after having been on display at Columb Barracks, Mulllingar for some time), a third at Collins Barracks Museum, Cork; as of December 2011 another 2-pdr was being refurbished at the Base Workshops at Curragh Camp.

The 2-pdr at Cathal Brugha Barracks, Dublin, in May 2011. There are three other 2-pdrs surviving in Ireland, one at Curragh Camp, one at the museum at Collins Barracks, Cork, and one being refurbished in late 2011 at the Combined Vehicle Base Workshops. (Colin Stone)

The 2-pdr at the Curragh, on its cruciform platform, devoid of its wheels, in a November 2011 photograph. (Colin Stone)

The Collins Barracks, Cork 2-pdr, in fairly good condition despite being displayed in the open, in a November 2011 photograph. Although not readily apparent in the photograph, the tires have different tread patterns. (Colin Stone)

A 2-pdr in the process of being refurbished at Curragh Camp Base Workshop, still lacking its definitive paint scheme, in December 2011. (Colin Stone)

Ordnance QF 6-pdr anti-tank gun

One 6-pdr belonging to the Artillery Corps is on display at Curragh Camp, and another is preserved at Cathal Brugha Barracks, Dublin.

The 6-pdr serving as a gate guardian at Curragh Camp, May 2009. The 6-pdr was in service for a relatively brief period of about ten years in the post-war Irish Army. (Ralph Riccio)

A 6-pdr anti-tank gun at Cathal Brugha Barracks, 31 August 1998, at a display marking the 50th anniversary of the "End of The Emergency." A profusion of handles, levers, and towing rings have been painted in red to dress up the gun for display. (Denis McCarthy)

The 6-pdr at Cathal Brugha Barracks, Dublin, in May 2011. Its appearance is quite in contrast to that of the surviving 6-pdr at the Curragh, which is in a more correct monochrome livery. (Colin Stone)

Ordnance QF 17-pdr anti-tank gun

Two 17-pdrs have been restored by the Irish Army and are on display at Cathal Brugha Barracks, Dublin. A third17-pdr has been restored and is maintained by the Cavan & Leitrim Railway Museum. Another 17-pdr was in the process of being refurbished at the Base Workshops, Curragh Camp, in December 2011. Two are also on display at Stevens Barracks, Kilkenny.

One of the 17-pdrs at Cathal Brugha Barracks, Dublin, in May 2011. There are at least five surviving 17-pdrs in Ireland: two at Cathal Brugha, two at Collins Barracks, Cork, and one at Dromod, County Leitrim, which is restored, and is now on permanent loan to the Cavan & Leitrim Railway Museum from the Irish Minister for Defence. (Colin Stone)

The second of two 17-pdrs maintained at Cathal Brugha Barracks. The barrel, which measured almost 14 feet in length enhanced its muzzle velocity, adding to its ability to penetrate heavy armor at long ranges. (Colin Stone)

The 17-pdr in the late stage of refurbishment, Base Workshop, Curragh Camp in December 2011. Although not evident in this photograph, the left wheel and tire are damaged and remain to be repaired, probably prior to painting in the definitive colour. (Colin Stone)

Brandt AM 50 120mm mortar

At least one AM 50 mortar has survived as a museum piece, located at Columb Barracks Museum, Mullingar.

A 120mm Brandt mortar stored at Columb Barracks Museum, Mullingar, September 2011. This particular mortar was deployed with the 35th Battalion in the Congo, where it fired some 225 rounds in support of the Swedish and Indian UN troop contingents. (Colin Stone)

QF 40mm Mk III (40mm Bofors L/60) anti-aircraft gun

A 40mm Bofors L/60 AA gun is on display at National Museum of Ireland, Collins Barracks, Dublin, and at least two others have survived, one at Cathal Brugha Barracks, Dublin, and another, in rather deteriorated condition, located at the Artillery School, the Curragh in 2011.

A Bofors L60 in rather worn condition, taken in April 2011 at the Artillery School, the Curragh. This piece may be refurbished in the future and be used as a gate guardian or museum exhibit. (Colin Stone)

The restored and preserved 40mm Bofors L60 anti-aircraft gun displayed at the National Museum of Ireland, Collins Barracks, Dublin. The large black and white image in the background shows the Bofors ready for action, manned by Irish gunners. (Colin Stone)

QF 3.7 inch anti-aircraft gun

One restored 3.7 inch anti-aircraft gun in the static mounting configuration is displayed at Curragh Camp.

The sole surviving Irish 3.7 inch heavy anti-aircraft gun photographed at Curragh Camp on 27 May 2009. This example has been nicely preserved and presents a rather imposing appearance. The 3.7 inch guns in the Irish inventory were both mobile and fixed versions; this is one of the latter. (Ralph Riccio)

QF 12 pounder 12 cwt coast defence gun

Collins Barracks, Dublin, has a very nicely restored OQF 12-pdr Mk I L coast defence gun on display. Additional 12-pdrs are statically mounted at Columb Barracks, Mullingar, and at the naval base at Haulbowline. A battery of four 12-pdrs is maintained at Fort Davis, used as salute guns on ceremonial occasions, and another four-gun salute battery, shown here, is located on East Pier, Dun Laoghaire. One ex-Ireland 12-pdr is on display at Newhaven Fort, Sussex, England.

QF 4.7 inch gun Mk IV

One 4.7 inch Mk IV coast defence gun (Serial number 1603) is displayed at the Fort Dunree Heritage Museum, County Donegal.

Three of the four 12-pdr salute guns at Dun Laoghaire are visible in this photograph taken in September, 2009. The guns are covered by protective tarpaulins to shield them from the seaside elements that have visibly corroded the iron railings on and leading to the gun parapet. (Roy Kinsella)

Detail including serial number of 12-pdr at Columb Barracks, Mullingar. (Colin Stone)

A 12-pdr at Columb Barrracks Military Museum, Mullingar. The museum had an extensive display of artillery related photographs and artifacts. (Colin Stone)

A very well restored 12-pdr, part of the National Museum of Ireland exhibit at Colllins Barracks, Dublin, December 2011. (Colin Stone)

A 12-pdr 12 cwt gun at the naval base at Haulbowline, September 2011. (Colin Stone)

BL 6 inch Mk VII naval gun

Two 6 inch Mk VII coast defence guns have survived as museum pieces at Fort Davis (ex-Fort Carlisle) in Cork Harbour; another four 6 inch guns are located at Fort Dunree in Lough Swilly, and two are in a state of deterioration, open to the elements, at Lonehort Battery, Bere Island. Two 6 inch guns – one of which is fully restored – remain in a casemates on Spike Island. Two of the guns at Dunree are guns original to the fort, and the other two were formerly located at Fort Shannon, in the Shannon Estuary, and were moved to Dunree in 1986. All of the guns displayed in the open at Fort Dunree are painted with a red oxide paint to provide some degree of protection against the elements. In addition, one 6 inch gun formerly located at Duncannon Fort, Co. Westmeath, is now part of a display at Newhaven Fort, Sussex, England, and has been very nicely restored.

A 6 inch coast defence gun photographed at Fort Davis on 21 July 1987. Despite the three-tone dazzle paint scheme remaining from its operational days, there are signs of rust on the gun shield. Since then, lack of care and maintenance have caused the guns to further deteriorate. (Denis McCarthy)

One of the two 6 inch guns at Fort Mitchel on Spike Island. One of the guns has been immaculately restored, while this gun shows its age, exposure to the elements, and general neglect, although it is largely intact and may be restored at some time in the future. (Gabriel Whitson)

One of the 6 inch gun positions at Fort Dunree. In common with all of the coast defence guns, the Dunree guns were sited with commanding views of the areas they protected, in this case, Lough Swilly. The gun is on the eastern shore of the loch, oriented toward the west. (David Magee, Fort Dunree Museum)

The two 6 inch guns that constituted the armament at Fort Shannon, in the Shannon Estuary, are now on display at Fort Dunree, on Lough Swilly. (David Magee, Fort Dunree Museum)

The ex-Ireland 6 inch gun being placed in its new home at Newhaven Fort, Sussex. (Philip Baldock)

The nicely refurbished 6 inch gun at Newhaven Fort, restored to fully operational condition subsequent to its return from Ireland. (Philip Baldock)

The 6 inch gun that had served at Duncannon fort, Waterford, Ireland, upon arrival in England, where it would ultimately be displayed at Newhaven Fort in Sussex after being completely refurbished. (Philip Baldock)

Appendix 4: Surviving Gun Towers

ACMAT VLRA

An undetermined number, but probably at least four, ACMATs are maintained by private collectors or groups in Ireland.

A very impressively preserved and maintained ACMT VLRA shown at Durrow Castle in April 2009. This is one of several ACMATs owned by private collectors or vehicle preservation groups in Ireland. The owner of this vehicle, Sean O'Sullivan, also has several other ex-Irish Army vehicles in excellent condition, including a Panhard VTT M3 armored personnel carrier, that he regularly exhibits at shows in Ireland. (Sean O'Sullivan)

AEC Matador

An AEC Matador (ZL 1257) is preserved at the National Transport Museum in Howth. It has been restored to its configuration as a military recovery vehicle, fitted with a Harvey Frost crane. It is painted in the grey livery appropriate for an Irish Army vehicle of its vintage.

The AEC Matador is a featured vehicle in the National Transport Museum, Howth, collection. ZL 1257 was converted as a recovery vehicle for the Army, and has been restored to that configuration, fitted with a Harvey Frost crane and a canvas shelter behind the cab to protect the crew. It is in the grey livery appropriate to the period of its service with the Army. (National Transport Museum, via Michael Corcoran)

Bedford MFR 2

An MFR 2 is preserved at the National Transport Museum, Howth. It is believed that at least one other ex-Irish MFR 2 is maintained by a private collector.

The Bedford MFR 2 8335 ZH belonging to the National Transport Museum is one of several ex-Irish Army MFR 2s that have survived, and is a Diesel engine version. The twin rear wheels are characteristic of the Bedford MFR2 in Irish service. (National Transport Museum, via Michael Corcoran)

Bedford MFR 2 8335 ZH at the National Transport Museum storage depot at Castleruddery in April 2010, showing effects of exposure to the elements. (Colin Stone)

Bedford RL

A Bedford RL is preserved at the National Transport Museum, Howth. It is believed that at least one other ex-Irish RL is maintained by a private collector.

The National Transport Museum's Bedford RL, HZH 155, minus its body. While in service, it was fitted as a Signal Corps communications van. The bright chromework surrounding the grille was retained on only a portion of the Irish Army's RLs, the remainder being painted in solid grey livery. (National Transport Museum, via Michael Corcoran)

Bedford RL HZH 155, in 2003, at Castleruddery. The vehicle has since been at least partially restored as part of the National Transport Museum collection. (Colin Stone)

Land Rover

Several ex-Irish service Land Rover variants are maintained, both by the Defence Forces as well as by private collectors in Ireland.

A 109 Series Land Rover in storage at Curragh Camp in May 2009. Despite being housed under cover, the paint is flaking badly. Some 350 Series III 109s were purchased by the Irish Army between 1972 and 1982. (Roy Kinsella)

Morris-Commercial C8 4 x 4 FAT

As of early 2010 a Morris C8 FAT, ex-Irish service, was awaiting restoration at the Cavan & Leitrim Railway Museum in Dromod, County Leitrim. The National Transport Museum, Howth, also has a C8 badly in need of restoration.

The Morris-Commercial Quad artillery tractor belonging to the National Transport Museum, Howth. This vehicle was used in an Italian film, for which the Museum was never paid. After filming, the Quad, having suffered body damage, was abandoned by the side of the road, where it was found by some students The effects of being stored in the open, plus probable vandalism, including the indignity of having graffiti applied, have taken their toll on the vehicle. (Colin Stone)

Morris-Commercial CDSW 6 x 4 artillery tractor

As of early 2010 a Morris CDSW was awaiting restoration at the Cavan & Leitrim Railway Museum in Dromod, County Leitrim. Another ex-Irish CDSW has been fully restored by a group in the United Kingdom.

Morris-Commercial CDSW ZC 5835, 30 July 2009. This CDSW, rescued from a Dublin scrap yard in the 1970s by a UK group that subsequently restored it, was driven to the Glen of Imaal by members of the group on the occasion of the last 25-pdr shoot in Ireland. It is shown here parked on the hardstand beside "Keogh's House." (Capt Ian Stewart)

Universal Carrier MK I/Mk II

The Irish Army maintains a Ford T16 in lieu of a Universal Carrier at Curragh Camp; a Universal Carrier (ZG 3469) is on display at the National Museum of Ireland, Collins Barracks, Dublin. A third carrier has been salvaged by a private collector in Clonmel, but is not in running condition.

The Ford T16 maintained by the Cavalry Corps. This particular carrier is not one of the Universal Carriers used in Irish service, but was bought by the Cavalry Corps as a representative vehicle of the same class. (Irish Defence Forces)

Appendix 5: Irish Artillery Barracks

Cathal Brugha Barracks, Rathmines, Dublin: Built by the British between 1810 and 1815, Cathal Brugha Barracks was originally named Portobello Barracks after the area of Rathmines (Dublin) in which it was located. Portobello was originally designed as a cavalry barracks, and the first British unit to occupy the barracks was the 6th Dragoon Guards, as well later by fusilier and light infantry units, among the Royal Irish Rifles. On 17 May 1922 the barracks were taken over by National Army troops, and it became the headquarters of the National Army under Michael Collins. In 1956 the installation was renamed after Cathal Brugha, one of the leaders of the 1916 uprising against the British, was Chief of Staff of the Irish Republican Army from October 1917 to March 1919, was the Chairman of the First Dáil Éireann; on 7 July 1922 Brugha, acting as Commandant of the Anti-Treaty forces defending O'Connell Street in Dublin, was killed. As of 2012 Cathal Brugha serves as Headquarters, 2nd Eastern Brigade, as well as hosting a number of combat arms and support units, including the 2nd Air Defence Battery, and the Military Archives.

A Honda Deauville 700cc of the Blue Hussars, the Presidential Escort detachment of the Army's 2nd Cavalry Squadron, which is based at Cathal Brugha Barracks, shown in front of HQ Eastern Brigade. (Colin Stone)

The square at Portobello Barracks – now Cathal Brugha Barracks – in the late 1800s or early 1900s. The British troops appear to be armed with Lee-Metford rifles. (NLI LROY 8903, courtesy of The National Library of Ireland)

The parade square at Victoria Barracks (now Collins Barracks), Cork, late 19th century. The Cork barracks had the largest such parade square in Europe. (NLI LCAB 02795, courtesy of The National Library of Ireland)

The Royal Barracks (now Collins Barracks), Dublin, in an image dating to sometime around the beginning of the 20th century. (NLI LCAB 8205, courtesy of The National Library of Ireland)

Collins Barracks, Cork: Collins Barracks, Cork, designed by John Gibson, were built between 1801 and 1806. Initially known simply as The Barracks, in 1849 the barracks complex was expanded and renamed Victoria Barracks to honour a visit made by Queen Victoria. The barracks has the distinction of having the largest military barracks parade square in Europe. In 1922 the barracks were handed over to the Irish State, and soon renamed for Michael Collins, the first commander in chief of the National Army and a native of County Cork. As of 2012 Collins Barracks serves as Headquarters of 1 Southern Brigade, as well as hosting two artillery batteries – No. 1 Gun Battery, 1st Field Artillery Regiment, and No. 1 Gun Battery, 31 Reserve Artillery Regiment, among other units.

Collins Barracks, Dublin: The oldest military barracks in Ireland, as well as the oldest inhabited barracks in Europe, Collins Barracks was built beginning in 1704 and expanded in the late 18th and early 19th centuries. The complex, whose buildings were built of Wicklow and Dublin stone with roofs made of Carnarvon slate, ultimately consisted of several large squares (Horse Square, Cavalry Square, Royal Square, Palatine Square and Brunswick Square), each open on the south side, and was originally known as The Barracks, and later became known as the Royal Barracks; through the 19th century up to 1,500 troops – primarily Regiments of Foot – were stationed at the barracks. During the 1916 Easter Rising, British forces were deployed from the barracks to quell the Irish insurgents on Usher's Island, the Four Courts, and the GPO on Sackville Street. On 17

December 1922 the barracks were handed over to the Irish Free State and shortly thereafter were named Collins Barracks (in Irish, Dun Uí Choileáin) after Michael Collins, who had been killed earlier that year. The squares, understandably enough, were renamed: Horse Square became Supply and Transport Square, Cavalry Square became Connolly Square – named for James Connolly, one of the leaders of the 1916 Easter Rising, executed by the British on 12 May 1916 – Royal Square became Collins Square, Brunswick Square became Pearse Square, and Palatine Square became Clarke Square, named for Tom Clarke, the Irish revolutionary leader who was, like Connolly, one of the leading forces of the 1916 Easter Rising and who was executed at Kilmainham Jail by the British on 3 May 1916. For about 70 years the barracks housed various commands and elements of the Free State Army and the Irish Army, including the 6th and 7th Field Artillery Regiments and 41 and 42 Batteries. The decision to close Collins Barracks as a military installation was made in December 1988, and in 1993 it was agreed that the complex should be developed for use by the National Museum of Ireland, but the barracks continued to be occupied by the 5th Infantry Battalion until it marched out of the barracks on 18 April 1997. On 18 September 1997 the inaugural exhibitions of the National Museum of Ireland, Decorative Arts and History opened in the portion of the barracks around the Clarke Square.

Exterior view of Collins Barracks in December 2011, looking from Pearse Square. Compared to the earlier vintage photograph, the most apparent difference in the building façade is that the stone exterior of the upper storeys has been covered over with a stucco finish. (Colin Stone)

The Palatine (now Clarke) parade square at Collins Barracks as it appeared in December 2011. This aspect of the barracks has changed little since it was built. (Colin Stone)

A battery of 60-pdrs with limbers beside them lined up for training at Collins Barracks, Dublin in 1942. (Shay Stone via Colin Stone)

Columb Barracks, Mullingar: The site for the barracks at Mullingar was purchased by the British War Department in 1807; work on construction began in 1814, and the barracks were first occupied in 1819. The original barracks, consisting of classically proportioned multiple-bay three storey barracks buildings, were built to house 1000 soldiers. Originally named Wellington Barracks for the Irish born Arthur Wellesley, 1st Duke of Wellington, in 1922 the barracks were renamed Columb Barracks in honor of Patrick Columb, a member of the National Army who was killed on Mary Street in Mullingar in 1922. From 17 February 1943 to 30 March 2012, Columb Barracks were the home of the 4th Field Artillery Regiment, and from 1 October 2005 to 30 March 2012 of 54 Reserve Artillery Regiment as well. Columb Barracks was scheduled to be closed on 30 March 2012, and its units transferred to Custume Barracks, Athlone.

An imposing scene of Wellington Barracks (now Columb Barracks), Mullingar during the early part of the 20th century, (NLI LROY 8098, courtesy of The National Library of Ireland)

A vintage postcard of Military Road in Mullingar, probably early 20th century. (Mullingar Barracks Museum via Colin Stone).

Sentries at the entrance to Mullingar's Wellington Barracks, as they were then known, probably early 20th century. (Mullingar Barracks Museum via Colin Stone).

Columb Barracks, Mullingar, home of 4th FAR from 1943 until 2012, with elements of 4th FAR formed up, date unknown. (Mullingar Barracks Museum via Colin Stone)

Connolly Barracks, Curragh Camp: Connolly Barracks is one of the many former British barracks located in Curragh Camp; until the barracks were turned over to the Irish State on 16 May 1922, the barracks were known as Stewart Barracks. After being taken over by the Irish, the barracks were renamed for James Connolly, Commandant of the Dublin Brigade during the 1916 Easter Rising, who was captured by the British and, even though badly wounded during the fighting, was executed shortly thereafter. As of 2012, Connolly Barracks are the home of No. 2 and No. 3 Batteries, 62 Reserve Artillery Regiment.

A part of Stewart Barracks, Curragh Camp, in the early 1900s. Stewart Barracks were renamed Connolly Barracks after the Irish State took over Curragh Camp from the British in 1922. (NLI EAS 2448, courtesy of The National Library of Ireland)

The statue of Saint Barbara, patroness of the Artillery, at Connolly Barracks, Curragh Camp, December 2012. The buildings to the left belong to Artillery School Air Defence and HQ 1st Air Defence Regiment. (Colin Stone)

Renmore Barracks, Galway City. Now named Dun Ui Mhaoiliosa (Mellows' Fort), these barracks were an important artillery barracks during The Emergency, but no longer have any artillery units stationed there. (NLI LCAB 4248, courtesy of The National Library of Ireland)

Dun Uí Mhaoiliosa, Renmore, Galway: Known as Renmore Barracks under British rule, it was home to the Connaught Rangers. The Irish name is Dun Uí Mhaoiliosa (Mellows' Fort), named for Liam Mellows, who participated in the Easter Rebellion in Galway and who later became one of the leading figures of the anti-Treaty faction; he was captured at the Four Courts and imprisoned until he was executed on 8 December 1922. Dun Uí Mhaoiliosa Barracks were the home of the 5th Field Artillery Regiment during The Emergency, and also when the regiment was reconstituted as an FCA unit from October 1959 until 2005 when it merged with the 9th Field Artillery Regiment to form 54 Reserve Artillery Regiment. The barracks complex is located east of Galway City proper and of Lough Atalia, at the southern end of Renmore Road, and no longer houses any artillery unit but now (2012) houses the 51 Reserve Infantry Battalion and 54 Reserve Engineer Company.

Magee Barracks, Kildare: Construction work on the Royal Artillery Barracks, Kildare, began in 1900; the site of the barracks was a farm known as Broadhook Farm, which was also the site of the Lock Hospital, built in 1868 and which remained open until about 1888. The barracks, known simply as Kildare Barracks, were first occupied in April 1902 by the 31st and 33rd Brigades, Royal Field Artillery. On 15 April 1922 the barracks were handed over to the National Army, and on 25 April the barracks became the centre for training of the new Civic Police force – later the Garda Siochána – that was to replace the Royal Irish Constabulary and was the location of the

The military barracks in Kildare Town, in the early 1900s. Renamed Magee Barracks some years after they had been taken over from the British in 1922, they are now (2012) in a state of great disrepair. (by kind permission of Mark McLoughlin)

The Royal Field Artillery Barracks, Kildare. Despite the fact that the barracks were taken over by the Irish State in 1922, they were not renamed as Magee Barracks until thirty years later, in 1952. (NLI EAS 2570, courtesy of The National Library of Ireland)

The entrance gate to Magee Barracks; the unusual cannon motif of the metalwork reflects the primary function of the barracks in their heyday. (Colin Stone)

"Kildare Mutiny" in May 1922. In March 1925 No. 1 and No. 2 Batteries of the Artillery Corps arrived at the barracks, and the barracks served as the headquarters of the Artillery Corps. In 1938 the first purpose built barracks built by the Irish State replaced the hutted lines of the Kildare barracks, and the new barracks were renamed Magee Barracks (the Irish is Dun Mhic Aiodh) in October 1952, after Gunner James Magee who had famously handled the six pounder at the Battle of Ballinamuck in September 1798. In 1954 a statue of Saint Barbara, patron saint of the Artillery, was unveiled at the barracks, but upon closure of Magee the statue was moved to the Artillery School and Depot at the Curragh. Magee Barracks were home to the 6th Field Artillery Regiment headquarters, and headquarters of the 1st Anti-aircraft Regiment. Due to reorgansation of the army, Magee Barracks were closed in September 1998. As of early 2012, the barracks were still the property of the Department of Defence but had deteriorated badly and were frequently subjected to arson and vandalism.

The desolate remains of the accommodations blocks at Magee Barracks, December 2011. The buildings have been completely vandalized; doors and windows are either missing or open, leaving the structures completely exposed to the elements. (Colin Stone)

The distinctively turreted Marlborough Barracks (Now McKee Barracks) in the Cabra area of Dublin. (NLI LCAB 6128, courtesy of The National Library of Ireland)

McCann Barracks, Templemore: McCann Barracks were originally named Richmond Barracks after the part of Templemore in which they were located. The barracks occupy the most elevated position in town, bounded by Barrack Street on the north and Thurles Road on the west, on land donated to the British crown by the local landlord, Sir John Craven Carden. During World War One, Richmond Barracks were used as a prisoner of war camp, housing more than 2,300 German and Austro-Hungarian prisoners. The barracks were turned over to the Irish State in 1922 and renamed McCann Barracks after Pierce McCann, a founding member of Sinn Fein in 1905, a member of Dáil Èireann and who died in prison in 1919. McCann Barracks has been the national training centre for the Garda Siochána since 1964 as well as home to the 1st Field Artillery Regiment and 31 Reserve Artillery Regiment.

McKee Barracks, Dublin: McKee Barracks on Blackhorse Avenue, were built by the British between 1888 and 1892 in the Dublin suburb of Cabra. The barracks were originally named Marlborough Barracks for the Duke of Marlborough. The National Army's first two artillery pieces – two 18-pdr field guns – were collected by four National Army officers from Marlborough Barracks on 27 June 1922, marking a milestone in the history of the Irish Artillery Corps. Later in 1922 the barracks were handed over to the Irish State, and on 10 July 1923 the 1st Field Battery was formed at the barracks. In 1926 the name was changed to McKee Barracks after Richard McKee, born in Finglas, and captured by the British on Bloody Sunday, 21 November 1920; McKee was later shot in Dublin Castle while trying to escape. McKee has continued to host Dublin-based artillery units and is the home of the 2nd Field Artillery Regiment and 62 Reserve Artillery Regiment.

Richmond Barracks, Templemore (now McCann Barracks) in a vintage photograph. (NLI LROY 8019, courtesy of The National Library of Ireland)

Murphy Barracks, Ballincollig: Murphy Barracks, which began life as the Royal Artillery Barracks, were built between 1806 and 1815 by the British Board of Ordnance to provide security for the Royal Gunpowder Mills at Ballicollig. The main square, Barrack Square, comprising four buildings, was laid out in 1811. Entrances to the barracks were the Main Gate, East Gate, and West Gate. Between 1875 and 1922, the barracks were expanded by the addition of the Officers Mess, the Officers Stables and the Carriage Store. The entire complex totaled some 431 acres. During the First World War and the War of Independence, the barracks housed British Army garrisons. In 1922 the barracks were turned over to the Irish State, and in 1940 the barracks were reopened as Murphy Barracks, remaining as an Irish Army artillery barracks until their closure in 1998, when 1st and 8th FAR vacated the premises. In 2003, 97 acres of the installation were purchased by developers, and is now the site of the Ballincollig Shopping Centre; some of the original buildings and walls have been preserved. Another plot of 27 acres was turned over to the Cork City Council for community use, and smaller parcels have been sold or turned over to the Department of Education and Science and to the Office of Public Works.

Sarsfield Barracks, Limerick: Sarsfield Barracks was built around 1795, at which time it was known as New Barracks. It was occupied by the British until 1922 when it was turned over to the Irish State. Sarsfield Barracks are named for Patrick Sarsfield, an Irish Jacobite officer. A somewhat interesting historical sidelight connected to Sarsfield is that in the 1870s a California National Guard unit, composed largely of ethnic Irish soldiers, was named the Sarsfield Grenadier Guards in his honor; its tradition is carried on by the 184th Infantry Regiment. Sarsfield Barracks are home to Headquarters, Southern Brigade Reserve Defence Forces and, among other units, are the home of the 3rd Air Defence Battery.

Military Barracks, Summerhill, Nenagh: Nenagh Military Barracks were built in 1832 and garrisoned by British regiments almost continuously until 1922 when, with the departure of the North Staffordshire Regiment, they were turned over to the Irish State. They were then occupied by the Irish Army for two years, and then by various government departments. The barracks are located on the eastern side of the road formerly known as The Old Turnpike, now known as Ormond Street. Although as of 2012 most of the property is in great disrepair, No. 3 Gun Battery, 31 Reserve Artillery Regiment is quartered there in prefabricated buildings.

Slievenamon Road, Thurles, Co. Tipperary: No. 2 Gun Battery, 31 Reserve Artillery Regiment, is located in what is believed to have formerly been a church building on Slievenamon Road, Thurles.

An aerial view of Murphy Barracks, Ballincollig, prior to being closed down in 1998. The barracks complex fell into great disrepair, but has since been rehabilitated into a shopping centre. (Sgt Mick Royanne, 1 FAR, via Colin Stone)

Bibliography

Bishop Chris, *The Encyclopedia of Weapons of World War II*. Barnes & Noble Books, 1998.

Carroll John, Davies Peter, *The Complete Book of Tractors and Trucks*. New York: Hermes House, Anness Publishing, Inc., 2002.

Chamberlain Peter, Gander Terry, *Anti-Aircraft Guns*. WW2 Fact Files. London: MacDonald and Janes's, 1975.

Chamberlain Peter, Gander Terry, *Anti-Tank Weapons*. WW2 Fact Files. London: MacDonald and Janes's, 1974.

Chamberlain Peter, Gander Terry, *Infantry, Mountain, and Airborne Guns*. WW2 Fact Files. London: MacDonald and Janes's, 1973.

Chamberlain Peter, Gander Terry, *Light and Medium Field Artillery*. WW2 Fact Files. London: MacDonald and Janes's, 1975.

Chamberlain Peter, Gander Terry, *Mortars and Rockets*. WW2 Fact Files. London: MacDonald and Janes's, 1975.

Clarke Dale, *British Artillery 1914-19*. Botley, Oxford: Osprey Publishing, 2004.

Coogan Tim Pat, Morrison George, *The Irish Civil War*. London, England: Weidenfeld & Nicholson, 1998.

Cottrell Peter, *The Irish Civil War 1922-23*. Botley, Oxford: Osprey Publishing, 2008.

Doyle Tom, *The Civil War in Kerry*. Cork: Mercier Press, 2008.

Duggan John P. *A History of the Irish Army*. Goldenbridge, Dublin: Gill and Macmillan, 1991.

Dunlevy Mairead, *Dublin Barracks/A Brief History of Collins Barracks, Dublin*. Dublin: National Museum of Ireland, 2002.

English Adrian J. *Irish Army Orders of Battle, 1923-2004*. Takoma Park, Maryland: Tiger Lily Publications LLC, 2005.

Fletcher David, *Universal Carrier 1936-48: The 'Bren Gun Carrier' Story*. Botley, Oxford: Osprey Publishing, 2005.

Green Ted, *History of Drogheda*.

Harrington Niall C. *Kerry Landing, August 1922*. Dublin: Anvil Books, Limited, 1992.

Haskew Michael E. *Artillery From the Civil War to the Present Day*. New York: Metro Books (for Amber Books), 2008.

Henry Chris, *The 25-pounder Field Gun 1939-72*. Botley, Oxford: Osprey Publishing, 2002.

Henry Chris, *British Anti-tank Artillery 1939-45*. Botley, Oxford: Osprey Publishing, 2004.

Hogg Ian, *Twentieth Century Artillery*. London: Amber Books Ltd., 2000.

Holmes Richard, *Bir Hakim: Desert Citadel*. New York: Ballantine Books Inc., 1971.

Hopkinson Michael, *Green Against Green: The Irish Civil War*. Dublin: Gill & Macmillan Ltd., 2004.

MacCarron Donal, *Step Together! Ireland's Emergency Army 1939-46 As Told By Its Veterans*. Dublin: Irish Academic Press, 1997.

MacCarron Donal, *Silent Sentinels. The Irish Treaty Forts*. Dublin: Nonsuch Publishing, 2008.

Martin Karl, *Irish Army Vehicles. Transport & Armour Since 1922*. Dublin: Cahill Printers, Ltd., 2002.

Moriarty Michael, *An Irish Soldier's Diaries*. Cork: Mercier Press, 2010.

Neeson Eoin, *The Civil War 1922-23*. Dublin: Poolbeg Press, Ltd., 1995.

Norris John, *Infantry Mortars of World War II*. Botley, Oxford: Osprey Publishing, 2002.

Riccio Ralph, *AFVs in Irish Service Since 1922*. Petersfield, Hampshire: Mushroom Model Publications, 2010.

Walsh Paul V. *The Role of Armoured Fighting Vehicles in the Irish Civil War, 1922-1923*. Paper presented to The Civil War Conference, Cathal Brugha Barracks, Dublin, 11-13 September 1997.

Younger Calton, *Ireland's Civil War*. London and Glasgow: Collins Clear-Type Press, 1970.
Dawson Comdt J.E., and Lawlor Lt C. *Coast Defence Artillery*. An Cosantoir, November 1973.
Donnelly Comdt.W. *Origins of 2 Field Artillery Regiment: 1943-1948*. Defence Forces Review 2009. Dublin: Defence Forces Printing Press, 2009.
Hudson John M. and Thompson F. Glenn, *The Artillery Corps 1936*. An Cosantoir, November 1973.
Kavanagh Comdt. P.D. *The Artillery School 1931-73*. An Cosantoir, November 1973.
Mulcahy Lt. Gen. P.A. (Rtd). *At the Beginning...* An Cosantoir, November 1973.
Nagle, Sgt David, *120mm Mortar Back in Action*. An Cosantoir, February 2009.
O'Donnell Comdt. M.J. *The Guns*. An Cosantoir, November 1973.
O'Halloran Comdt. C.P. *The Ack-Ack*. An Cosantoir, November 1973.
O'Leary Comdt R.P. *The Foundation of the Artillery School and A Brief History of Its Development Between 1931 and 1998*. Article published in Artillery Corps Anniversary Book 1923-1998.
O'Sullivan Terry. *Thoughts of Depression*. An Cosantoir, November 1973.
The Irish Times, 26 June 1967
The Irish Times, 29 June 1967
The Irish Times, 23 July 1981
Artillery Corps 1923-1998, November 1998. 75th Anniversary Commemorative Booklet, edited by Captain T.M. Clonan.
3rd Field Artillery Regiment 1959-1980: 21 Years of Loyal and Dedicated Service, edited by Capt. F. Hannon. Freshford, Co. Kilkenny: Wellbrook Press, 1980.
Collected reminiscences, 7th Field Artillery Regiment, 1942-1992, Shay Stone, with forward by M.D. Shannon Colonel, Director of Artillery
Soldiers and Chiefs: The Irish at War at Home and Abroad From 1550. Dublin: National Museum of Ireland, 2007.
Lonehort Battery, Bere Island, West Cork. Report to the Bere Island Projects Group on the tourism potential of the Lonehort Battery and Five Year Development Plan 2008-2002. Donaldson Tourism and Heritage Consultancy, 20 October 2008.
Unsourced paper, *The Four Courts Guns*. Military Archives artillery folder.
Camp Diary, Coolmoney Camp, 18th May 1936 to 5th December 1941.
Defence Force Regulations. Manual of the Mortar (81m.m. and 3-inch). Training Regulations No. 30. (Revised Edition, 1946). Dublin: Printed and published by the Stationery Office. 1946.
Manual, Ordnance QF 18-pdrs Marks 1, 2, 2* – Land Service 1940
Manual, Ordnance QF 18-pdr, IV, IVA, IVB – Land Service, 1940